Paradoxes
Guiding Forces in Mathematical Exploration

Paradoxes

Guiding Forces in Mathematical Exploration

Hamza E. Alsamraee

This page is intentionally left blank

First edition published September 2020

Printed on acid-free paper

Typeset using LaTeX

Edited by Rick Keys

ISBN 978-1-7357156-1-2 (paperback)
ISBN 978-1-7357156-0-5 (ebook)

LCCN 2020917274

10 9 8 7 6 5 4 3 2 1

www.hamzaalsamraee.com

To the DailyMath community.

CONTENTS

AUTHOR'S NOTE

Hi, I'm Hamza. I'm a teenager into all sorts of things, with math, science, and philosophy being at the very top! So, I figured I would write a couple of books about paradoxes, the first centered around mathematical paradoxes, and the second around scientific paradoxes. Paradoxes do not only contain beautiful math and science — they have an entire rich background of philosophy behind them. This book was bred out of my fascination for these paradoxes, but also out of a need to share these wonderful entities with the world.

I'm used to sharing mathematics and science with the world by now. I was always one to explain everything I learned to friends and family, and soon enough I began freelance tutoring to save some money for my wants. I then got a job at an after-school educational company and began explaining away concepts, from simple addition to advanced calculus, to children of all different ages. I loved it.

I wanted to grow out these efforts a bit, however. Therefore, I made my math page, DailyMath, about a year ago. I started out by posting my favorite mathematical creatures — integrals. I'd post challenging integrals and put up solutions for them everyday. The page surprisingly gained traction, and I slowly began building a community. On the side, I was writing my first book — *Advanced Calculus Explored*. By this point, I had been writing this book for a few years, and I even used some of its problems in the early days of DailyMath.

I published that book in December of 2019. Writing it was extremely rewarding, but seeing people reading it was even more so. Surprisingly, the book went on to be a #1 Amazon bestseller, and in multiple categories! My first book solidified my goals to be an educator — so I became determined to keep curating content, writing books, and spreading knowledge. The DailyMath community kept growing, and so did my commitment to being an educator. DailyMath, alongside its sister pages, is now a community of over 100,000 people from 140 different countries.

Truly, this book would not have been possible without that community of enthusiasts, students, and professionals. For those in the community already, thank you, and for those just joining, welcome! I sincerely hope you enjoy this book.

Part I

INTRODUCTION

1

WHY MATHEMATICAL PARADOXES MATTER

> **Paradox**
> *noun*
> A seemingly absurd or self-contradictory statement or proposition that when investigated or explained may prove to be well founded or true.
>
> — From Lexico.

Surely a succinct definition, but is it the only one? Let's look into it in more depth. Paradox comes from the combination of the Greek prefix para- (meaning "beyond" or "outside of") and the verb dokein ("to think"). Outside our thinking. Herein lies a crucial aspect of paradoxes: they, through some mechanism, escape our beliefs. Nonetheless, paradoxes are not simply riddles or parlour tricks, but a fundamental building block of mathematics.

There are arguably many types of "paradoxes," each of a specific nature. The renowned American philosopher Willard Van Orman Quine classified paradoxes into three categories: the veridical, the falsidical, and the antinomical. For the purposes of this book, we will use his classification.

The veridical paradox is the closest to the definition of paradox mentioned above. A veridical paradox seems absurd, even contradictory, but upon further inspection turns out to be true.

Veridical paradoxes make for great brain-teasers as well as amazing educational material.

Let us divert for a moment to the world of theater. *The Pirates of Penzance* was a comic opera by the Victorian-era duo William Gilbert and Arthur Sullivan which still enjoys much attention today. The story revolves around Fredric, who at 21 becomes an indentured servant to a group of pirates. His indenture[1] specified that he remain apprenticed to the pirates until his 21st birthday. But wait — why isn't he released yet?

In fact, not only has he not been released, but he has 63 years left to serve on his contract! Did the pirates simply break the contract's rules? What's behind all of this?

It turns out that Fredric was born on February 29. With our current calendar, he only has his birthday on leap years — which happen every four years[2]. As of the time Fredric is 21, he has had only five birthdays! So his 21st birthday won't be until $21 * 4 - 21 = 63$ years. Yikes, always read the small print!

The story is much richer than just outlined[3], as is the paradox. We just completely resolved the veridical paradox, but our next step stems from the need of understanding why it is the way it is. Why do we even have leap years? Why do leap years occur at the frequency they do? There will never be a shortage of questions and directions we can take after resolving this paradox. We can delve into astronomical observations, historical context, and beyond. In analyzing and resolving this paradox, we are able to learn much more than just its resolution, and effectively get more bang for our buck. For this reason, these types of paradoxes, which include the likes of Hilbert's hotel and Schrodinger's cat, excel as educational adventures.

A falsidical paradox, on the other hand, is essentially a fallacious argument that somehow passes through our logic filter

1 Historically, an indenture refers to a legal contract that specified terms for indentured servitude.

2 More precisely, a leap year happens every four years except on years divisible by 100 but not by 400.

3 I certainly recommend watching the 1983 film adaptation.

just fine. These paradoxes seem technically sound, but end up in contradictions and absurdities due to subtle errors in reasoning. Take for example the following argument. For two non-zero numbers a and b, we can do some algebraic manipulation to obtain the result that $2 = 1$.

$$a = b \tag{1}$$

$$a^2 = ab \tag{2}$$

$$a^2 - b^2 = ab - b^2 \tag{3}$$

$$(a + b)(a - b) = b(a - b) \tag{4}$$

$$a + b = b \tag{5}$$

$$2b = b \tag{6}$$

$$2 = 1 \tag{7}$$

This is absurd! But how did we get there? Pause for a minute to see if you can spot the error in the proof. The logic certainly seems sound, but upon further inspection we realize that we divided by $a - b$ in (4). Since $a = b$ was our starting claim, $a - b$ has to equal zero. We committed the sin of dividing by zero! In fact, we divided zero by zero in (4), as zero multiplied by any number is still equal to zero. Of course, this is a huge flaw in the logic of the argument.

Why can't we divide by zero? Well, let's think of how we define division. Division in its most elementary form is the inverse of multiplication. For example, asking what $\frac{10}{2}$ is equal to is the same as asking "What number multiplied by 2 gives 10?" This might seem mundane, unnecessary even, but bear with me. We will find that in many of the examples considered in this book, that single ignored mundane step in the logic proves itself to be crucial. Now, let's apply our basic logic behind division to the fraction $\frac{0}{0}$. What number, multiplied by zero, gives zero?

Aha! One multiplied by zero equals zero. After all, any number divided by itself should equal one, right? But wait — two multiplied by zero also equals zero. Zero multiplied by zero equals zero. And the list goes on, infinitely. Any number multiplied by zero, by the standard axioms of arithmetic, is equal to zero[4].

Siri, the "virtual assistant" found in most Apple devices, has a lovely answer:

> Imagine that you have zero cookies, and you split them evenly among zero friends. How many cookies does each person get? See? It doesn't make sense. And Cookie Monster is sad that there are no cookies, and you are sad that you have no friends.

Ouch. Despite Siri's rudeness, it's right — $\frac{0}{0}$ does not make any sense! Nonetheless, we can still learn a lot from exploring the paradox a bit more. The example of dividing by zero touches on very fundamental ideas about arithmetic and mathematics at large. Paradoxes like these provoke curiosity, and they can easily be a segue into more advanced concepts.

My little brother, much like I was when I was his age, loves a good round of arithmetic questions. He has a pretty good handle on numbers; adding, subtracting, multiplying, and dividing with much accuracy. After getting bored with constantly asking him arithmetic questions, I decided to throw him a trick one to keep him busy for a while: *what's 1 divided by 0?*

He was startled and eventually gave up. I asked him to ask Siri, after which he became even more curious[5]. I continued, asking him to ask Siri what zero divided by zero was. Even more curious now, and a bit offended, he became very eager to have an understanding of why. The next hour saw me introduce him to not only the aforementioned explanation, but the concept

4 Axioms are statements which are so self-evident they are taken as true without proof. They serve as the starting point for all further mathematics (i.e. proofs). The word comes from the Greek axioma: 'that which commends itself as evident.' We will see their importance in later chapters, especially in antinomical paradoxes.

5 Siri's response to this was "Please don't make me divide by zero. That would be like me asking you to grow a third arm."

of limits. He loved it, and only ended up being more inquisitive about what's out there.

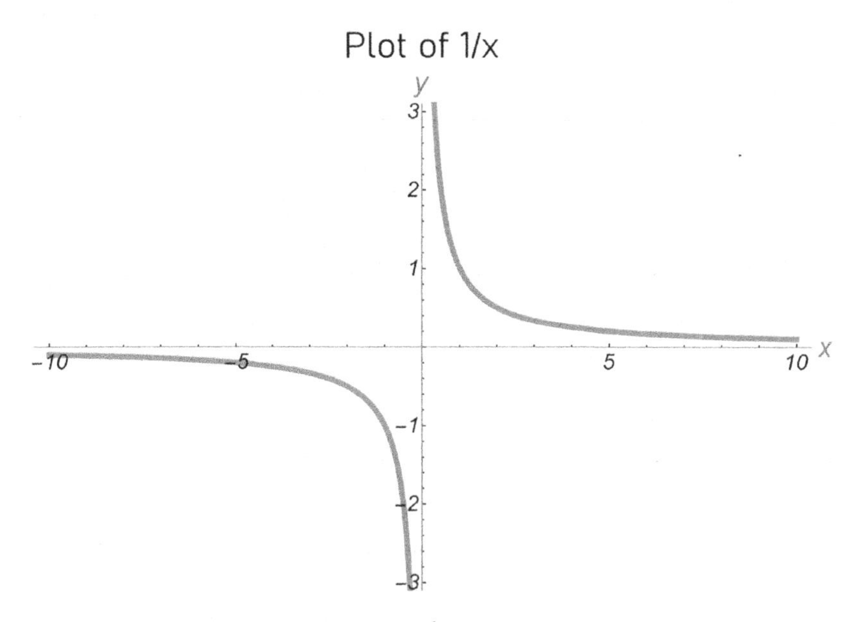

Figure 1: While the graph of $y = \frac{1}{x}$ approaches $-\infty$ from the left of $x = 0$, it approaches $+\infty$ from the right. Therefore, the limit does not exist, and the expression $\frac{1}{0}$ is considered undefined.

Now, my seven year old brother can't do calculus, but the introduction to these higher concepts is never an exercise in futility. Beyond the entertainment one gets from delving into such paradoxes, falsidical paradoxes are extremely thought provoking, and through this thought, one cannot help but develop a more rigorous understanding of the field under study. By inviting students to look at the argument that $1 = 2$ and dissect it, they will not only be exercising their mathematical knowledge but much more: their problem-solving and critical thinking skills. The skill of dissecting these fallacious arguments extends beyond the scope of mathematics and the fields of STEM at large. These are the very skills that must be taught in an age where humans' role as mathematical computers is virtually non-existent, while their role as analytic problem-solvers is only growing. Perhaps the type of paradox that most uses these analytic skills are antinomical paradoxes.

Antinomical paradoxes: my personal favorite. More commonly, these are known as antinomies. Antinomies invite us to question frameworks of reason that are considered very much grounded, but are, or seem to be, mutually incompatible. Antinomies compel the mathematician to revise these frameworks of reason, mainly through considering the foundational ideas of the field. These paradoxes are arguably the most fundamental and the biggest instigators of change in mathematics. A famous example is given by Russell's paradox, which revolutionized set theory into the rigorous field it is today[6].

There is also a great deal of educational value to be found in antinomies. Through understanding paradoxes like these, students gain perspective into the fields they are studying. By understanding how concepts and ideas are related in a much larger framework and actively thinking about these frameworks of reason, one gains a much richer understanding of the topic. Antinomies like Russell's paradox are already widely used in the introduction of many courses at the undergraduate level. And many of them are the subject of much discussion and even research in the mathematical community to this day. The framework in which mathematics and many other fields are encapsulated is constantly evolving, and antinomical paradoxes bring insight into how and why.

While each type of paradox is in its way a valuable complement to any course of study, paradoxes as a whole have been shown to promote critical thinking through various studies. We'll touch on some of the psychology behind this.

The confusion that comes when encountering paradoxes like the ones surveyed in this book, the feeling of uncertainty and suspicion of core ideas, can be described as *cognitive conflict*.

---

6 More on that in the next part!

This state of cognitive conflict is central to Jean Piaget's theory of learning.

Jean Piaget was a Swiss psychologist who made major contributions to the psychology behind child development, especially in regards to education. He was the second most highly cited psychologist of the 20th century. As a researcher and Director of the International Bureau of Education, he was a key figure in advancing childhood education.

Figure 2: Jean Piaget (1896 - 1980)

Piaget's ideas regarding education are still widely taught and are used for the development of educational systems to this day, so we ought to take a look at them.

According to Piaget's learning theory, an individual constructs knowledge when they receive an input from their environment and "assimilates" that input into their current "mental structures." If newly assimilated information conflicts with any existing mental structure, however, the result is considered to be a disequilibrium. Cognitive conflict can then be regarded as a state of disequilibrium, fueled by conflict either between cognitive structures and experience, or between various existing cognitive structures. In our study of paradoxes, veridical and falsidical paradoxes are much closer to the former than the latter, while antinomical paradoxes embody the latter. In Piaget's theory, this disequilibrium pushes and motivates an individual to resolve the conflict and attain a new state of cognitive "equilibrium" through incorporating new information. Regaining this equilibrium or "cognitive harmony" results in what Piaget names accommodation. It is this process that builds new structures.

To be useful educationally, however, these cognitive conflicts require careful construction. The purpose of introducing these cognitive conflicts is not to confuse the student, but to motivate the search of a resolution. Some studies that have replicated this, thereby strongly substantiating Piaget's theories, were done by Irwin (1997) and Fujii (1987)[7]. An overwhelming body of evidence has repeatedly shown that the application of Piaget's framework makes for more intriguing and thought-provoking learning at all levels. It is here where we see the value of paradoxes — they are the ultimate cognitive conflict.

Another framework to consider the value of paradoxes through is the educational value of *counter-examples*. Personally, I was always a fan! My linear algebra professor would always include a large number of "True or False" questions in his tests to confirm his students had properly understood the concepts at play. The first test of this sort was a nightmare. The statements seemed entirely reasonable initially, yet many had plenty of exceptions. The result was a lot of low scores, and a lot of learning after the test. Seemingly simple questions incited countless discussions which enriched the whole class's understanding of the topic at hand, and tied fundamental concepts into a proper framework of understanding.

The reason behind the success of these counter-examples is that they do not require any prelearned algorithm or formula to solve. The thinking is much more free, and therefore produces much more value. It engages not only the direct chapter material, but other knowledge and mechanisms of thought. Let's take a wild turn here and look at a far tangent: weightlifting. Yes, it in fact explains this phenomenon beautifully!

In effect, this approach to learning versus the algorithmic approach is similar to the concept of a *compound* exercise and

7 References are all included at the end of the book, as well as other works I used for writing this book!

an *isolation* exercise. While compound exercises like the squat can engage many muscles, concepts in our scenario, isolation exercises like a bicep curl only target one muscle, or one concept.

Figure 3: An example of a compound exercise and an isolation exercise.

A more extreme form of an isolation exercise is done when it is done on a machine. For example, instead of doing dumbbell bicep curls, one might want to try out a bicep curl machine. This results in an even more restricted targeting of muscles, and even though this can be effective for focusing on single muscles, it is not your optimal exercise if you are looking to save time and still get a good workout in!

Now, that bicep curl machine — or in our scenario a narrow problem with only one way to go about it — is bound to get you results. But, that exercise should never be the main focus of a normal training regiment! In observing that, we realize that a proper mathematics education should never focus on very narrow problems which you solve with formulaic algorithms. Instead, it should be much more free and flexible, and focused more on conceptual understanding.

Mathematician Sergiy Klymchuk offers many examples of such "counterexamples" in his book *Counterexamples in Calculus*, as well as an interesting account of using paradoxes in the classroom. Calculus is riddled with counter-examples. Take for example this question:

> If a certain function, $f(x)$, is continuous for all real numbers x, could it also be differentiable nowhere?

At first, the answer seems like an obvious no! But, as it turns out, there exist functions that are continuous everywhere but differentiable nowhere: fractal functions. The first such function to be discovered was the Weierstrass function.

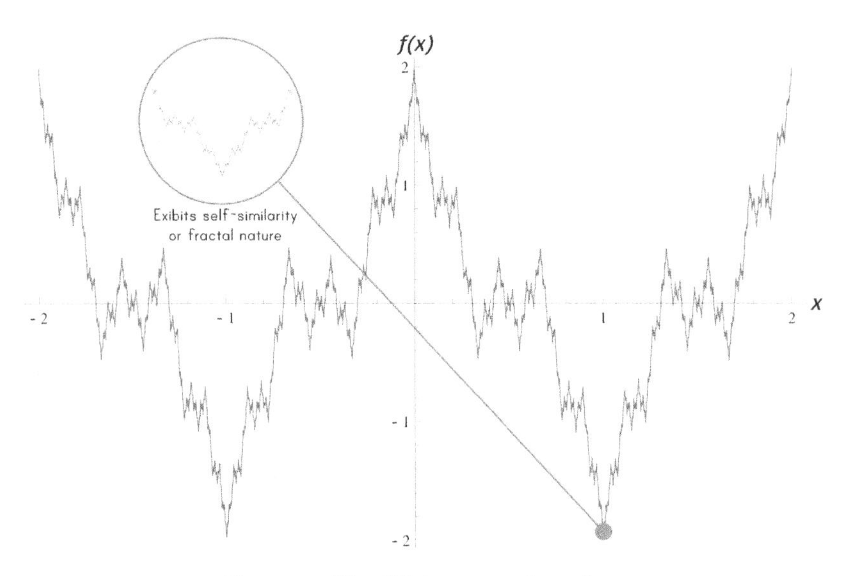

Figure 4: The Weierstrass function.

The fact that few students thought the answer was yes did not stem from any explicit teaching material, but from the subconscious thoughts that arise from only being introduced to "nice" or "well-behaved" functions. As a consequence, many students form incorrect assumptions. Mathematics educators Anne Watson and John Mason argue for this in their book *Mathematics as a Constructive Activity*, stating:

> *Deliberate searching for counter-examples seems an obvious way to understand and appreciate conjectures and properties more deeply. Such a search could be within the current example space or could promote extension beyond ... In our view, learners will inevitably encounter nonexamples of concepts and counterexamples to conjectures if they are actively exploring and constructing their own spaces.*

In the article "Using counter-examples in teaching and learning of calculus: Students' attitudes and performance," Klymchuk presents the results of a survey of over 600 students across ten countries. Out of these 600+ students, 92% reported that they thought the use of counter-examples was effective! What might this have to do with paradoxes? Well, counter-examples are arguably best embodied by paradoxes. To give a quick example, Russell's paradox will prove to be a counter-example to the notion that "any definable set exists."

Paradoxes, which can often be thought of as counter-examples on steroids, attack our most fundamental conceptions of truth and show us our logical inconsistencies. In doing so, paradoxes build a stronger base for understanding and strengthen analytical ability. By navigating through paradoxes, we are essentially navigating through unknown territory, armed only with our sense of logic. This is a tool that develops mathematical and analytical ability more than any other.

It is always entertaining as well as highly educational to observe how people react to paradoxes. As the admin of the popular mathematics page *@daily_math_*, I witnessed this first hand with many of the paradoxes I post[8]. Responses range from astonishment to confusion to outright denial, and each of these responses show why we need more paradoxes in mathematics education. By frequently challenging students' intuition, their understanding of mathematics becomes much more solid.

---

8 More on my observations later.

1.1 A SHIFT IN MATHEMATICS EDUCATION?

In an age where we rely more and more on computation, we need to reform our educational system to teach analytical skills that machines cannot replicate.

Conrad Wolfram, CEO of Wolfram Research Europe, has outlined a plan to do this in his book, *The Math(s) Fix: An Educational Blueprint for the AI Age.* Wolfram distinguishes mathematical skill from the ability to compute mathematical expressions efficiently: mathematics is much more. He argues for a more computational-based educational system, and with that comes a new focus — not only on the implementation of computational methods — but on broader mathematical and analytical skills as well.

By reforming mathematics education to be more focused on fundamental conceptual understanding instead of rather laborious computation, much better mathematicians, scientists, engineers, and problem-solvers in general emerge from our educational institutions. If we can accept that mathematics is part of a larger skill to think analytically, incorporating paradoxes into mathematics education is simply the next logical step. In the first chapter of *The Math(s) Fix*, Wolfram outlines two different conceptions of mathematics. While one is seen as the backbone of our society — driving innovation in industry, technology, healthcare, and government — the other math is seen as an unnecessary requirement imposed on many, which few will ever apply "in the real world." The way culture, especially in countries like the US, views mathematics is a paradox of sorts — on one hand, it is revered, on the other, it is hated.

And that is exactly why incorporating the paradoxes outlined in this book, as well as countless possible others, is crucial to the advancement of mathematics education globally. If we as a society wish to advance students to become the next generation of mathematically-fluent problem-solvers, we must teach skills that aren't bound to be much better performed by a computer.

We ought to expand from a focus on algorithmic methods done by hand to a focus on conceptual understanding, which is where paradoxes can be crucial to discuss. The move towards a more conceptual focus in mathematics education is already in motion; but paradoxes are a key educational tool that seems to be missing from the discussion.

The dilemma of hand-computation in mathematics education, fortunately, lessens as one climbs up the difficulty ladder of mathematics. An introductory calculus course will include much more hand-computation than a real analysis[9] course, which is much more focused on rigor, logic, and very little on computation. And even when computation is needed, only a minority is done by hand.

This rigor presents another problem: it effectively becomes the new computation, again overwhelming many students. Rigor is important, especially in a real analysis course, but a course entirely on rigorous methods without any perspective on why these methods exist is ineffective. While mastering the methods is a crucial aspect of learning any higher mathematics course, understanding the methods' importance puts everything into perspective. By connecting the dots through a rich understanding of the *why* behind these methods — methods like epsilon-delta proofs, bijections, etc. — a student benefits more from the study of a particular course.

Instead of introducing a real analysis course, let's say, with a rigorous study of epsilon-delta proofs, why not start it with a discussion of why $.999\cdots = 1$? This paradox will naturally lend itself to discussions of the real number line and its construction, limits, amongst other important mathematical concepts. By probing through these types of fundamental paradoxes, students at all levels develop a deeper understanding of the material at

9 Real analysis is a subfield of mathematics that deals with the rigorous study of the behavior of real numbers, real-number sequences and series, as well as real functions. It can be thought of as the rigorous foundation which elementary calculus rests upon.

hand as well as context and perspective. Another advantage that introducing paradoxes in mathematics education lends itself to is the interactive nature that the paradoxes necessarily foster. By introducing the real analysis class to the $.999\cdots = 1$ paradox, and letting them probe it collaboratively, students facilitate learning within each other and understand various perspectives on the problem. Unlike solidified mathematical methods, which contain very little room for a student's own thought, the discussion of paradoxes encompasses much more than one correct view or solution.

Moreover, an undeniable fact about paradoxes is that they are intellectually exciting. By introducing them as puzzles to be discussed, the barrier of fear in most students tends to disappear. A problem like:

> Prove that $\lim_{x\to 0} \frac{\sin x}{x} = 1$ using the epsilon-delta definition of the limit.

Has a very cut-and-dry, rigorous solution. However, a problem like:

> Discuss why $.999\cdots = 1$.

is much less intimidating than the epsilon-delta proof. It also offers much more insight into the nature of limits, numbers, calculus, etc. This is because paradoxes are often in the basis of many concepts, and understanding them gives us much more than the truth value of their statements. Another vital characteristic of this problem is that it is open-ended! Just like those problems that occur in real-life.

Now, mastering epsilon-delta proofs and other methods is crucial to develop a proper understanding of real analysis. But so is understanding the underlying reason they exist and the larger framework in which they were developed. Through studying paradox, the student can develop a much greater appreciation of why the methods they learn in the classroom exist, thereby strengthening mathematical ability.

1.2 THE PARADOX AS AN ILLUSION

Take a look at figure 5. Is square A darker than square B? There's many ways to go about answering this question. We can:

Figure 5: Checker Shadow Illusion

- Open this image in some editing software and use the eyedropper tool to measure the color of the pixels on both square A and square B.
- Use a photometer.
- Cut the squares separately and see if they are the same shade.

- etc.

But wait, why all the fuss? Square A is obviously darker! The surprising illusion here is that the shades of square A and B are in fact identical. To see this, let's extend the color of square A to square B.

Figure 6: Extending square A onto square B

As it turns out, square A and B have the same colors! What's going on here? To get at the root of this, we'll have to uncover some psychology. The key concept here is that our brain is not a computer simply digesting raw data. In addition to digesting raw data, however, our brain uses previous experiences to interpret and make sense of this data to make for meaningful sensory signals.

As stated by Edward Adelson, Professor of Vision Science at MIT and creator of this illusion:

> *As with many so-called illusions, this effect really demonstrates the success rather than the failure of the visual system. The visual system is not very good at being a physical light meter, but that is not its purpose. The important task is to break the image information down into meaningful components, and thereby perceive the nature of the objects in view.*

As a matter of fact, these complex relationships between stimuli and perception are the basis of a whole field of study: psychophysics. Similar to how psychophysics shows us that our vision is not programmed to be a camera, but rather an active component of our interpretation of the world, we are not merely proof-generating algorithms. This is a positive, though. Unlike the computer proof-generating algorithm, we are capable of thinking "outside the box" to a much higher degree.

In essence, a part of this book is doing what we shall call *psychomathematics* — a field of study analyzing the relationship between one's mathematical beliefs and their reaction to fuzzy mathematical truths. Not mathematical psychology, which is the study of psychology mathematically, but rather the psychological study of humans learning and encountering mathematics. This can be considered as a subfield of the study of mathematics education as it is crucial to mathematics education research. This concept is already widely discussed, although without a fancy name like psychomathematics!

The visual effect we see in figure 6 resembles our intuitions in regards to mathematical truths. Through various underlying mechanisms, we decide if a tricky mathematical statement is true or not. In many instances, we aren't even aware exactly why! And here comes the double-edged sword that is intuition: it is amazing that the human brain is capable of such tasks, but sometimes this intuition carries us too far — into realms of inconsistencies and fallacies.

At the root of it, the role of intuition in examining paradoxes, mathematical or otherwise, tends to be due to very subtle yet fundamental beliefs about the structure of a field. For an example of intuition in action, we can look at the field of chemistry and *chemical intuition*. Chemists often have a great chemical intuition where they can form educated guesses on the properties of various compounds and elements. In the case of elements, the driving belief is periodic trends, but as the chemical substance in question gains complexity, the driving beliefs behind the educated guess tend to increase in subtlety. This is what we see with complicated mathematical paradoxes: often times, the underlying beliefs that cause someone to deny a particular mathematical statement due to its paradoxical nature are not easy to uncover at first. Paradoxes in the end remind us that our intuition can be deceiving, and in turn force us to construct a more holistic understanding of the subject at hand. A lovely case is given by Russell's paradox, discussed in chapter three.

1.2.1 *The Paradox as a Magic Trick*

Illusions come in many shapes, and some people make it their career's goal to use illusions to entertain the masses: magicians! A magician's purpose is to achieve a contradiction — to pull a rabbit out of an empty hat, to "levitate" objects despite gravity, to split a person in half then reassemble them without any harm. In many ways, our uncovering of paradoxes in this book will be very similar to a magician's study of magic tricks. No, magicians can't simply create rabbits out of nowhere or split people in half (if they do not want to go to prison, that is!). Upon further study of the mechanisms behind the magic tricks, we discover a set of events we were never aware of!

We then find out that there are no contradictions in the work of a magician, only *apparent* contradictions. When we dig deeper in mathematics to uncover the reason for the seeming contradictions, our understanding of the idea is reformed. We are now one layer deeper than we were previously, with a reformed lens through which we view mathematical structure. From that viewpoint, there is no contradiction, only apparent contradiction.

And in comparing the study of mathematical paradoxes to a magician studying magic tricks, we ought to be reminded of another aspect: entertainment! There have been countless books on riddles and puzzles, and the especially counterintuitive ones provoke much interest because of how entertaining they tend to be. In essence, the paradoxes in this book are riddles and puzzles of much higher complexity and with much more at stake: foundational ideas about mathematics. Ideas about the nature of the infinite, whether mathematics' is limited or not, and the interpretation of data, amongst countless other concepts.

Entertainment also reminds us of a key factor in why people do mathematics. Yes, mathematics can produce very practical results. From the physical sciences and engineering to the social sciences, mathematics continues to shape our society's intellectual progress and continually expands its impact. But that's not why professionals or hobbyists do mathematics. Sure, mathematics can give some very practical results, but that's not why we do it!

1.3 NEW MATHEMATICS

Mathematics, as beautifully as it describes the natural world, is demonstrably "invented" and adapted to our needs. The question of whether mathematics in its pure form is invented or discovered is not at stake here, it is simply the fact that *our* mathematics is invented. As recently as the 20th century, we reimagined the entire framework in which mathematics existed within[10]! Recognizing that *our version of mathematics* is invented gives a reason to develop a larger understanding of what "makes sense," which occasionally involves delving into philosophy. As we will see, by the 20th century, it was

10 More on this in chapters 4 and 5.

recognized that mathematics is simply the process of deducting truths and theorems from a set of axioms which form a certain mathematical structure. But, how could we know that this structure is useful and non-contradictory? This is where the study of paradoxes comes into the development of mathematics, as in many times where mathematics was charting unknown territory, paradoxes guided mathematical progress.

Grant Sanderson, popular mathematics communicator known through his YouTube channel's name, 3Blue1Brown, sheds light on this in his video "What does it feel like to invent math?" He suggests that the discovery of non-rigorous truths about nature and its mathematical properties drive humans to make rigorous mathematical theory, which is the "invention" I am referring to here. That rigorous mathematical theory continues to drive even more discoveries, thereby producing a never-ending cycle of mathematical evolution.

Regardless of whether you believe mathematics is a discovery or an invention, the mere fact that we can formulate an *alternate* mathematics or make alternate frameworks in which we look at mathematics — something that we will witness in the chapter on Russell's paradox — drives us to the same conclusion: paradoxes guide mathematical progress.

1.4 A GUIDE ON HOW TO USE THIS BOOK

This book contains paradoxes of all kinds. Given that, there are various degrees of mathematical and philosophical difficulty inherent to each paradox. Some paradoxes are especially hard to gain an understanding of, while some will perhaps come intuitively to some readers. Although I have tried to make each paradox understandable to the largest possible audience, some paradoxes will inevitably seem too difficult, pushing the reader to skip them. If that is the case for you, I encourage you to read through the introductory few pages of the chapter. These pages often contextualize the paradox through various means and provide an easy-to-understand summary of the essence of the paradox discussed in that chapter.

The book is divided into a few parts, one for each general topic of mathematical paradox. The first part is arguably the most influential on mathematics' history, but all the paradoxes discussed in this book have had a significant effect on the development of mathematics, either directly or indirectly.

If you are unfamiliar with the notational aspects of say, set theory, I encourage you to make use of the notation guide found at the end of this book. The appendix also provides background and essential knowledge regarding many fields. Even though this book uses some undergraduate level concepts at times, for example indefinite integrals, bijections, etc, the appendix is there to help the less experienced reader become familiar enough with these concepts to still understand the content of the book.

I encourage you to make full use of this book, beyond enjoying the beautiful mathematics presented. The ability to deal with paradox, and counter-examples in general, is a crucial analytical skill that can be carried into mathematical and non-mathematical fields alike. Moreover, as you think of the paradoxes presented throughout the book, try to learn more about yourself, your beliefs, and your inherent biases. Make sure to position them in a larger framework of your subconscious mathematical and philosophical beliefs. Enjoy!

2

PARADOXES IN SCIENCE

Paradox is by no means exclusive to mathematics. Any field of inquiry will inevitably run to paradox, and science is no exception. This is especially true in physics, where much of the progress in the 20th century was driven by the paradoxes of quantum mechanics and relativity. Perhaps best illuminated by Brian Greene, professor of physics at Columbia University and popular science communicator, in his book *The Elegant Universe*:

> *As with all apparent paradoxes ... under close examination these logical dilemmas resolve to reveal new insights into the workings of the universe.*

The paradoxes of physics that emerged in the 20th century proved to be a paradigm shift for all subsequent physics, giving us the two pillars of modern physics: quantum mechanics and general relativity. I hope to dedicate a new book to these paradoxes and other scientific ones such that both mathematical and scientific paradoxes are explored in depth!

Fortunately, science, especially physics, is much more lucky in mathematics in regard to its use of paradoxes in education. The theories of relativity and quantum mechanics are often introduced with the paradoxes that inspired them, thereby provoking much thought and providing the context that is extremely helpful in digesting these advanced ideas. Nonetheless, the second book shall serve as a unique, entertaining and thought-provoking survey of these paradoxes put into larger historical and philosophical perspective.

To get a taste of the type of paradoxes that will be introduced in the next book, we will take a very famous example and aim to dissect it. As it turns out, paradoxes do not only confound students and amateurs, but also trouble some of the greatest minds in science. Even Einstein fell prey to this because of his heavy bias against quantum theory! He tried throughout his life to construct thought experiments that in one way or another undermined the conclusions of quantum mechanics. Needless to say, he did not win this battle. Nonetheless, it will be fruitful to look at the story behind one of his well-documented attempts.

In the 1930s, Einstein and two other scientists — Boris Podolsky and Nathan Rosen — proposed an intriguing thought experiment. Dubbed the EPR paradox after the initials of the trio, Einstein believed this paradox to be a clear indication of an incompleteness in quantum theory — something he rejoiced in as an avid opponent of the theory. Specifically, in their paper *Can Quantum-Mechanical Description of Physical Reality be Considered Complete?*, the trio attempted to invalidate the current notion of entanglement by posing that it violates the theory of special relativity.

Before we get into the nitty-gritty of the EPR paradox, what is entanglement anyway?

Definition

Quantum entanglement is a state of particles in which their quantum states cannot be described independently, i.e. the system is entangled in some way. This is usually produced when particles interact or are generated in a specific manner.

To wrap your head around this, consider a pair of shoes. If someone were to put each shoe in a separate box while you are blindfolded, and reveal to you the identity of a shoe in one box — whether it is a left or right shoe — you immediately know the identity of the other shoe. The key difference here is being a left or right shoe is a "classical" state, while quantum states are what entangled pairs show correlation in. One such quantum state is *spin* — in essence an intrinsic form of angular momentum carried by particles like the electron. A particle can either have spin "up" or spin "down." By knowing the spin of one entangled particle, you immediately know the spin of the other(s) entangled with it.

Knowing what spin is, we can now attempt to understand EPR's thought experiment. Consider two particles formed from pure energy. By the principle of conservation of angular momentum, we know that if one particle is spin "up" then the other must be spin "down", and vice versa. However, this is only true if the particles are measured in the same direction. A spin measurement of "up" means that the spin of a particle is aligned in the direction of measurement, while a measurement of "down" means that it is opposite the direction of measurement.

But what if the direction of measurement is perpendicular to the spin? What if it is at some arbitrary angle from the spin direction, say 30 degrees? As it turns out, there is a formula for this. For an angle θ between the direction of spin and the axis of measurement,

$$P(\text{Up}) = \cos^2\left(\frac{\theta}{2}\right)$$

where $P(A)$ indicates the probability of event A occurring. So, it is no longer a guaranteed outcome what the measurement will be.

Einstein wondered if he could play a game of probability to violate the law of conservation of angular momentum. We will try to encapsulate his argument in the following thought experiment. Consider two entangled particles. Name one of our particles Alice, or particle "A", and the other particle Bob, or particle "B". Let's assume Alice has spin "up." Automatically, we know Bob has spin "down". If we measure these particles at an axis perpendicular to the spin, there is only a $\cos^2\left(\frac{90}{2}\right) = \frac{1}{2}$

chance of the particles yielding opposite spin outcomes and conserving angular momentum. But wait — that means 50% of cases do not conserve angular momentum.

This is outrageous! Is the law of conservation of angular momentum flawed? Well, not really. What is at fault here is the very act of assigning Alice and Bob predetermined spins. Quantum mechanics posits that Alice and Bob do not have well defined quantum states, with spin being one of these quantum states. These particles are in a "superposition" of quantum states in which they are *all* quantum states at once before one makes a measurement.

When entangled particles are measured in the same direction, physicists have found that the particles always have opposite spins regardless of the axis of measurement. In other words, the analogy of the shoe is of perfect relevance here. By knowing one particle is measured spin "up," the other is measured spin "down" regardless of how far separated these particles are.

Now, take a moment to ponder the inherent strangeness of this phenomenon! It seems as if though Alice and Bob, even if placed on opposite sides of the universe, could instantaneously communicate with each other. This is where we encounter Einstein's dilemma with entanglement, as faster than light communication is prohibited in his theory of special relativity. The speed of causality, more commonly described as the speed of light in a vacuum, is the universal speed limit that Einstein's theories would not be able to survive without. If faster than light communication is true, it destroys the entire validity of the theory of special relativity.

Therefore, Einstein and his peers objected and proposed a solution: Alice and Bob have had "hidden variables" all along which determine their respective measurements. These hidden variables contain information which encodes what Alice and Bob give out at each measurement, which is in some way hidden from us observers until a measurement is made. This suddenly resolves the faster than light communication dilemma, but is it true?

EPR's explanation was a direct attack at quantum theory's probabilistic framework — something Einstein was never a fan of — so we ought to be careful in accepting it. At the time, there seemed to be no way to resolve this. The concept of "hidden variables" was so vague and seemingly infinitely flexible, so it was hard to disprove the EPR conclusion.

That is, until John Stewart Bell.

John Stewart Bell was a Scots-Irish physicist who resolved the EPR paradox through **Bell's theorem** and made other lasting contributions to quantum mechanics. Besides his contributions to quantum mechanics, he primarily researched particle physics and particle accelerators at CERN.

Figure 7: John Stewart Bell (1928 - 1990)

Bell objected to Einstein's fundamental conflict with quantum mechanics and set out to prove his objection. He ingeniously developed a way to test Einstein's "hidden variable" theory through a pair of entangled particles. His experimental design goes something like this: Alice and Bob, the entangled particles, were to be sent to one of two detectors. Alice goes out to detector A while Bob goes out to detector B. Each detector is in one of three settings, with settings being equiangular from each other (i.e. 120 degrees away from each other). But how exactly can we rule out hidden variable theories from this experimental set-up?

Consider hidden variables as some sort of plan that the particles both consent to before being measured. The sole criterion this plan has to meet is if the two particles are measured

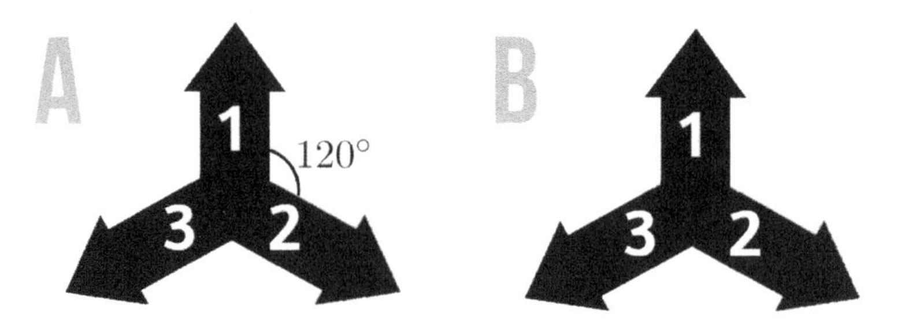

Figure 8: Symbolic representation of the settings of the two detectors, detector A and detector B. The numbered arrows represent the settings and their directions.

in the same direction, they must give opposite spins. From this criterion, we have two generic possible plans.

Alice could decide to give spin "up" in all directions while Bob agrees to give spin "down" in all directions. Another possible plan is for Alice to give spin "up" in the first setting, spin "down" in the second, and spin "up" in the third, with Bob giving the opposite measurements at each setting. All other plans are mathematically equivalent. In the first type of plan, the spins will be opposite 100% of the time, but in the second type of plan, the situation differs. We can make a table to keep track of all the measurements for the generic plan discussed above, keeping in mind that it is mathematically equivalent to all other possible plans.

By counting the number of times that the measurements were opposite and dividing it by the total number of instances, we get that the second plan should give opposite spins with a frequency of $\frac{5}{9} \approx 55.6\%$. Hence, if one combines the two plans, there should be opposite spins *at least* 55.6% of the time. However, when this experiment was carried out in 1972, it gave opposite spins only 50% of the time!

Incredibly, this is consistent with the laws and predictions of quantum mechanics. To demonstrate this, suppose that Alice

Measurement setting	B1	B2	B3
A1	⇑⇓	⇑⇑	⇑⇓
A2	⇓⇓	⇓⇑	⇓⇓
A3	⇑⇓	⇑⇑	⇑⇓

Table 1: The order of spins is Alice, Bob (can also be thought of as detector A, B).

is measured to be spin "up" at setting 1. We then immediately know that Bob is spin "down" if measured at setting 1. However, Bob goes through setting 1 only $\frac{1}{3}$ of the time, so we have to look at what happens with settings 2 and 3 to get the general probability. Both settings 2 and 3 make a 60 degree angle with Bob's spin. Therefore, Bob's spin should be measured as "up" $\cos^2\left(\frac{60}{2}\right) = \frac{3}{4}$ of the time it passes through settings 2 or 3. The overall probability that Bob is measured to have spin up is then $\frac{2}{3} \cdot \frac{3}{4} = \frac{1}{2}$ of the time, so Bob will be measured to have spin down $\frac{1}{2}$ of the time as well. This logic can be extended for any initial setting. This is in exact agreement with experiment! A Feynman quote is called for here:

> *A paradox is not a conflict within reality. It is a conflict between reality and your feeling of what reality should be like.*

Let's delve into why Einstein, a huge scientific figure and one of the most brilliant minds in physics, was so vehement about this paradox. As mentioned before, Einstein was in fervent opposition to quantum mechanics for much of his life. The quantum revolution of the 1920's brought this bias of Einstein's to the public spotlight, especially after the now-standard work of Max Born on the probabilistic nature of quantum mechanics. In a letter addressed to Born, Einstein writes:

> *Quantum mechanics is certainly imposing. But an inner voice tells me that it is not yet the real thing. The theory says a lot, but does not really bring us any closer to the secret of the 'old one'. I, at any rate, am convinced that He is not playing at dice.*

This quote is famously paraphrased as "God does not play dice." But, Einstein's attitude in this letter is much closer to one of healthy skepticism than of stern bias. It wasn't until the Fifth Solvay Conference of October 1927 that Einstein's skepticism turned into outright dismay.

The Fifth Solvay Conference is perhaps the most famous scientific conference in history. Figure 9 shows a famous picture from the conference that tends to circulate around social media as well as various other circles, which you may well have seen before!

Figure 9: Among the 29 attendees were Albert Einstein, Neils Bohr, Marie Curie, Max Planck, and many other renowned scientists.

Out of the 29 scientists who attended the conference, 17 of them went on to be Nobel prize winners. That's one successful bunch!

Neils Bohr was in this group. He was a prominent quantum theorist and one of the founders of the new field of study. He was even awarded a Nobel prize a year after Einstein for his work on atomic and quantum theory. Bohr, a firm believer in quantum mechanics, would engage with the skeptical Einstein in a series of scientific disputes after the conference. These debates are considered crucial because of their importance in shaping the philosophical framework which quantum mechanics rests upon.

Figure 10: Niels Bohr (left) with Albert Einstein (right) in 1925.

The scientific consensus remains that Bohr emerged victorious. The case of Einstein shows us that paradoxes can fool anyone, even Einstein. And that is why we ought to study their history so we do not repeat Einstein's mistakes.

Part II

FOUNDATIONAL PARADOXES

Pure mathematics is, in its way, the poetry of logical ideas.

— Albert Einstein

Key word: logic. But, for much of history, mathematics and logic were considered two distinct fields. Bertrand Russell, prominent logician and mathematician, discussed this strange fact in his *Introduction to Mathematical Philosophy*:

> *Mathematics and logic, historically speaking, have been entirely distinct studies. Mathematics has been connected with science, logic with Greek. But both have developed in modern times: logic has become more mathematical and mathematics has become more logical. The consequence is that it has now become wholly impossible to draw a line between the two; in fact, the two are one. They differ as boy and man: logic is the youth of mathematics and mathematics is the manhood of logic. This view is resented by logicians who, having spent their time in the study of classical texts, are incapable of following a piece of symbolic reasoning, and by mathematicians who have learnt a technique without troubling to inquire into its meaning or justification. Both types are now fortunately growing rarer. So much of modern mathematical work is obviously on the border-line of logic, so much of modern logic is symbolic and formal, that the very close relationship of logic and mathematics has become obvious to every instructed student. The proof of their identity is, of course, a matter of detail: starting with premises which would be universally admitted to belong to logic, and arriving by deduction at results which as obviously belong to mathematics, we find that there is no point at which a sharp line can be drawn, with logic to the left and mathematics to the right.*

As mathematics became much more rooted in logic, a move for its formalization emerged. In this part, we will discuss the various paradoxes and related developments that led to the formalization of mathematics. We will find that much of this part is philosophical in nature, and that is no coincidence. From the 2500-year-old paradoxes of Zeno and Parmenides to the modern paradoxes of Russell, these foundational paradoxes have revolutionized how we view mathematics and its philosophy.

3

ZENO'S PARADOXES

Perhaps the best way to begin this chapter is with the oldest of mathematical paradoxes: Zeno's paradoxes. These paradoxes were devised by the Greek philosopher Zeno of Elea almost 2500 years ago.

Zeno of Elea was an ancient Greek philosopher, perhaps best known for his paradoxes. Although none of his works remain intact today, he was referenced by many later important philosophers such as Aristotle and Plato. His arguments were one of the first examples of a method of proof called *reductio ad absurdum,* comparable to "proof by contradiction."

Figure 11: Zeno of Elea (c. 490–430 BC)

Zeno's paradoxes deal with some rather strange aspects of continuous space and time through examining motion. Motion is a concept which pervades both our day-to-day life as well as physics. Early physics could arguably be reduced to the study of it! However, Zeno thought that motion is a mere illusion.

An outlandish claim indeed. Let's get into some of Zeno's background before we delve into the paradoxes.

Zeno was a student of the Eleatic school of philosophy, which was founded by Parmenides of Elea. Parmenides was himself a very accomplished philosopher, often being credited as the father of metaphysics: a branch of philosophy examining the nature of reality. Parmenides was strongly opposed to any attempt aimed at acquiring knowledge through the senses. To him, our senses of sight, hearing, smell, taste, and touch were fundamentally flawed tools to understand the world. Instead, he advocated for a philosophy based entirely on logic. Parmenides rejected the senses and embraced the mind.

Without delving too much into why this philosophy is flawed at its heart[1], it shows how Zeno's intellectual upbringing as a student of Parmenides influenced his philosophical beliefs. The Eleatic school had a solid point: our sensory interpretation of the world is prone to much error. If anything, this book is in part a warning against relying wholly on sensory interpretation and intuition! But, Parmenides and his students might have taken their claims a little too far. We must remember that moderation is key.

Following in the spirit of his teacher, Zeno devised a set of paradoxes that he believed proved motion is a mere illusion. That is a paradox for sure. But where was his logic wrong? The paradoxes themselves proved much harder to dissect and resolve than the Eleatic school's philosophy, remaining a topic of discussion to this day — thousands of years after they were laid down by Zeno.

Why so? They raise interesting questions about space, time, and their fundamental nature — in a subtle way. At the heart of many of these paradoxes is some assumption regarding the

1 Without any sensory information, the mind and logic itself would have nothing to work on. Studying the arguments made by the Eleatic school shows that even though they dismiss sensory perception altogether, they use it to formulate their theories and philosophy. Perhaps that is the biggest contradiction, and the only real one, present in this chapter!

nature of the divisibility of space and/or time. Most of what we deal with in the "real world," i.e. matter, is not infinitely divisible. We know that all matter is made of atoms, and those atoms are made of tiny fundamental particles such as electrons. Mathematicians, on the other hand, operate on infinitely divisible quantities. One only needs to look at calculus to see infinities and infinitesimals everywhere. Take for example the concept of Riemann sums, which is the basis of the integral calculus we use to understand the world. For those familiar with calculus, this demonstration is all too familiar. For those not familiar or needing a refresher, here is a simple introduction. You can denote taking an integral, denoted by the symbol $\int$, of a function over a certain interval, say $x = a$ to $x = b$, as

$$\int_a^b f(x)\, \mathrm{d}x.$$

One can think of this integral as the "area under the curve $f(x)$ from $x = a$ to $x = b$." Suppose we would like to take the area under the curve $f(x) = x^2$ from $x = 0$ to $x = 1$. We can approximate this area using rectangles, as seen in figure 12. But, to get a more accurate number, we can increase the number of rectangles in our approximation, as seen in figure 13.

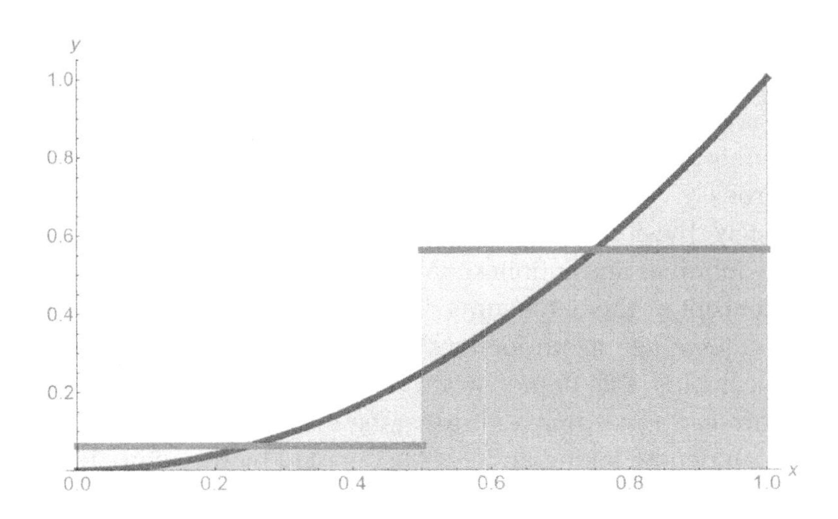

Figure 12: We can approximate the area under the curve with a set of rectangles. Figure is taken from my book *Advanced Calculus Explored*.

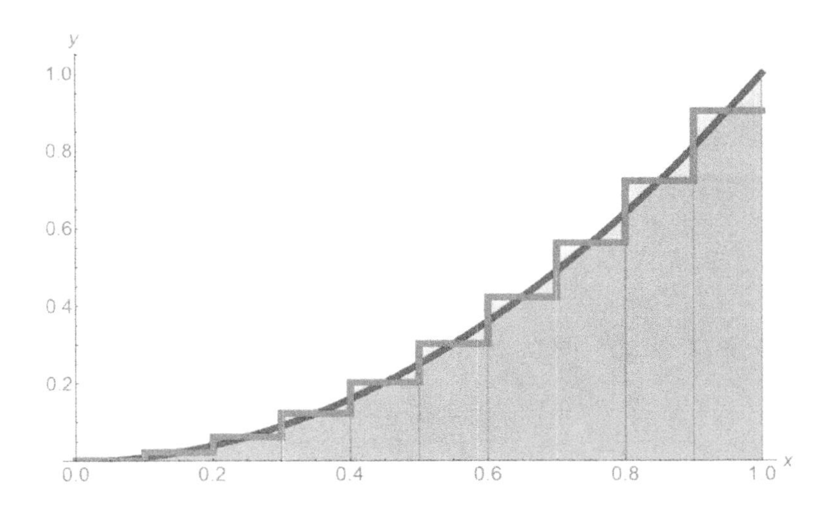

Figure 13: Now with much more rectangles approximating the area, we have a more accurate number for the integral. Figure is taken from my book *Advanced Calculus Explored*.

Ultimately, the integral is taken by taking the *limit* of the approximating area as we increase the number of rectangles to infinity. With this insight, we can go back to our original discussion.

In calculus, we treat the world as infinitely divisible. General relativity, which uses very advanced calculus, assumes spacetime is smooth and continuous, not discrete. However, in the real world — which we use calculus to describe — space and time aren't infinitely divisible. Or are they? We do not know — but we do know that our physics breaks down at extremely small scales. This disconnect is extremely interesting to investigate. Even though Zeno did not know about general relativity or calculus, we can see the same questions about the nature of space and time appearing in his work. We'll hopefully talk about this dilemma more in the next book of the series.

Nine of Zeno's paradoxes were preserved through Aristotle's *Physics*, but a few were essentially equivalent — being mere reformulations of each other. In this chapter, we will discuss the most prominent three: the Dichotomy, the Arrow, and Achilles and the tortoise. These paradoxes shaped mathematical understanding for centuries, and are described by many as fundamental to the development of mathematical thought. The influential British polymath Bertrand Russell described the paradoxes as "immeasurably subtle and profound." Russell went on, stating that:

> *Zeno's arguments, in some form, have afforded grounds for almost all theories of space and time and infinity which have been constructed from his time to our own.*

With these paradoxes' influence in mind, let's jump right in!

3.1 THE THREE PARADOXES

3.1.1 *The Dichotomy Argument*

Scenario

> *That which is in locomotion must arrive at the half-way stage before it arrives at the goal.*
>
> As recounted by Aristotle, Physics VI:9, 239b10

Suppose Zeno wants to get from point A to point B. Before he gets to point B, he must get halfway there. Let's call that point C. Before he gets to point C, however, he must get halfway there. Let's call that point D. If we keep this thought process forever, we end up with a sequence:

$$\left\{ \ldots, \frac{1}{4}, \frac{1}{2}, 1 \right\}$$

This, Zeno maintains, is impossible as there is an infinite amount of tasks to be done. Furthermore, this scenario presents a second problem: it contains no first distance to run. This is because any possible finite first distance could be divided, and therefore would not be "first." To Zeno, this meant the trip cannot even begin. The conclusion Zeno then drew was that travel over any finite distance can neither be completed nor begun, so all motion must be an illusion.

But, the conclusion that motion is an illusion is preposterous! Where has Zeno gone astray? I'll leave you to think about it independently before we uncover it in more depth together.

3.1.2 *The Arrow in Flight*

Scenario

> *If everything when it occupies an equal space is at rest, and if that which is in locomotion is always occupying such a space at any moment, the flying arrow is therefore motionless.*
>
> As recounted by Aristotle, Physics VI:9, 239b5

Imagine an arrow moving in space. At any duration-less instant of time, the arrow is neither moving to where it is nor to where it is not. We know that for motion to occur, an object must change its position. But in any instant of time the arrow is not changing its position. Since time is comprised entirely of instants, then motion is impossible.

It is noteworthy that unlike the other paradoxes in this chapter, which divide space, this paradox divides time. Also, this paradox does not divide time into segments but rather points.

3.1.3 *Achilles and the Tortoise*

Scenario

> *In a race, the quickest runner can never overtake the slowest, since the pursuer must first reach the point whence the pursued started, so that the slower must always hold a lead.*
>
> As recounted by Aristotle, Physics VI:9, 239b15

Suppose Achilles, a Trojan war hero, is in a race with a tortoise. Since the tortoise is slower, it gets a head start of 10 meters.

Apparently, Achilles never overtakes the tortoise! This is since whenever he moves, the tortoise also moves a bit more. The argument is similar to that of the Dichotomy.

Zeno was not attempting to make a point about infinity. As a member of the Eleatic school — which regarded motion as an illusion — he saw it as a mistake to suppose that Achilles could run at all. Subsequent thinkers, finding this solution unacceptable, debated for over two millennia on the exact falsities present in his argument.

3.2 SOME RESOLUTIONS

Close this book for a second. Now walk for a few seconds. Voila! All of these paradoxes are resolved. We did it — we proved motion is real. Apparently, this is what happened when Diogenes the Cynic heard Zeno's arguments. Upon hearing the claims, without saying anything, he stood up and walked in order to demonstrate the falsity of Zeno's conclusions. The conclusions are absurd, that's certain! But let's not get too ahead of ourselves — we will not be satisfied with merely demonstrating Zeno's conclusions as false. We need to dig into exactly why his reasoning is incorrect. As we saw in the introductory chapters, that is the most rewarding task when studying paradoxes.

3.2.1 *A Purely Mathematical Attempt*

When I first learned about Zeno's paradoxes, I had recurring thoughts of calculus and limits. I felt that these paradoxes could be easily resolved using the ideas of calculus because of how much they resemble the problems discussed in calculus' early history. As it turns out, Zeno's paradoxes — at core concerning infinite processes — indeed have a (mostly) ready mathematical explanation.

Infinite processes were theoretically problematic in calculus for centuries, mostly due to the lack of mathematical rigor early calculus had. Many calculus arguments were heuristic, as seen in the case of the great advances made by the mathematician Euler in the 18th century. However, by the late 19th century, the works of Weierstrass and Cauchy added much rigor to calculus and established a solid foundation of logic on which infinite processes can be studied. Through the rigorous basis of calculus, we can study Zeno's paradoxes with much more insight. In the following few pages we will discuss each paradox, with a mathematical way of resolving them.

Dichotomy

The Dichotomy can be understood of in terms of *series*. Series are simply infinite sums whose terms have some pattern behind them. They are wonderful mathematical objects that are as entertaining as they are useful. From Taylor series to Fourier series, the study of series has led to incredible tools for science and engineering and has given a new rigor to calculus.

Specifically, the Dichotomy can be resolved through looking at *geometric series*. A geometric series is one that has a *common ratio* between its terms, i.e. the pattern in the added numbers is a multiplier.

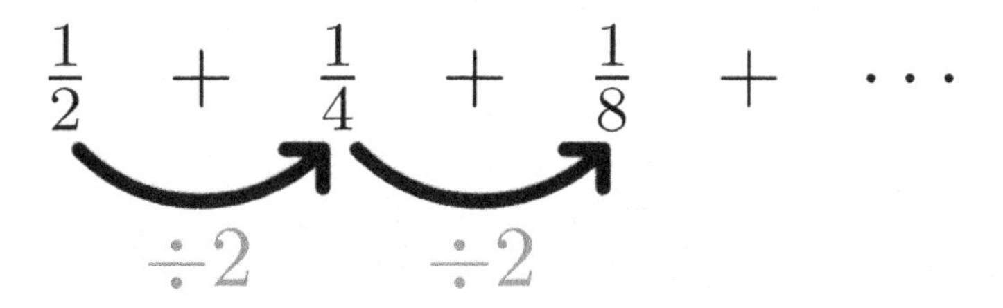

Figure 14: The geometric series found in the Dichotomy argument.

When looking at this in the viewpoint of limits and calculus, we can easily show that this sum equals 1. Denote the sum as S, then:

$$\begin{aligned} S &= \frac{1}{2} + \frac{1}{4} + \frac{1}{8} + \cdots \\ 2S &= 1 + \underbrace{\frac{1}{2} + \frac{1}{4} + \frac{1}{8} + \cdots}_{=S} \\ 2S &= 1 + S \\ S &= 1 \end{aligned}$$

For those of you who are new to this type of manipulation, I understand why you think this is too beautiful to be true! But it is, and we will look at some cases where this manipulation goes wrong in the part on infinity. Rigorously proving it is equal to one requires a bit more mathematical machinery, but this manipulation shall do for our purposes. This manipulation shows that an infinite sum of finite terms does not always diverge — something that Zeno did not realize. We can now see how the Dichotomy is a fallacious argument.

Not so fast! Are we really dismantling the logic behind Zeno's original argument, or merely propping up a strawman for our attack? If we delve into the original formulations of Zeno's dichotomy argument, we find that Zeno's problem is not with finding the sum of an infinite amount of elements but rather with finishing a task that has an infinite number of steps. As you probably noticed, there's no end to the summation we did above. How can we get from point A to point B if there is an infinite number of steps and/or events we need to traverse to get there? How is motion even possible then?

This is where the notion of *supertasks* comes into play! Supertasks are tasks which take an infinite number of steps but a finite amount of time to complete. The term supertask was coined by the philosopher James F. Thomson, who devised the Thomson lamp thought experiment. He was fully convinced that motion is not a supertask and that supertasks are impossible.

Consider this: we have a lamp that is programmed to turn on and off at specific times. At $t = 0$, it is off. It turns on at $t = \frac{1}{2}$. At $t = \frac{3}{4}$ it is turned off. At $t = \frac{7}{8}$ it is turned on again. The process continues for an infinite amount of steps.

$$0, \frac{1}{2}, \frac{3}{4}, \frac{7}{8}, \frac{15}{16}, \ldots$$

OFF ON OFF ON OFF

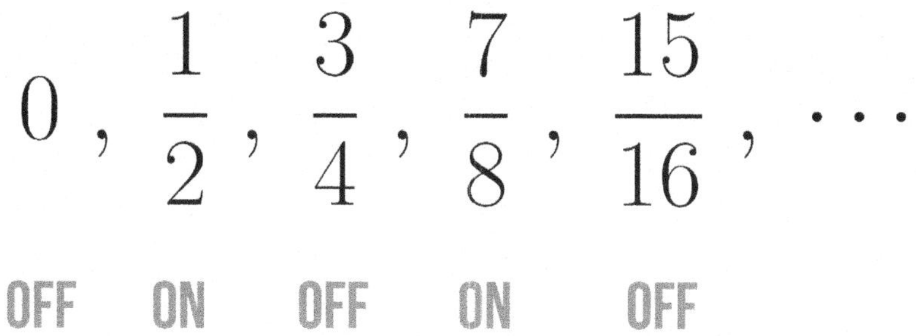

Figure 15: Being a supertask, this process is supposed to go on for an infinite amount of steps!

At $t = 1$, is the lamp on or off? Pause for a minute to think about it independently. To Thomson, there was no non-arbitrary method of attacking that question. He continues his argument and claims that this thought experiment produces a contradiction. The lamp cannot be on at the end of the sequence, for it will be shut off in an instant. On the other hand, it cannot be off at the end of the sequence, for it will be turned on in an instant. In many ways, the question of the lamp's status is similar to asking "Is the largest natural number even or odd?" It simply makes no sense to ask that question.

This leads us to the conclusion that the lamp is neither on nor off. Yet we stipulate that the lamp has to be either on or off — a contradiction.

We will touch on supertasks even more on the section on infinity, so it suffices to leave this here. Yes, I did leave this paradox with no full resolution. Philosophers to this day disagree if it, and Zeno's other paradoxes, have been fully resolved — maybe this is your time to shine!

The Arrow

Once again, taking a purely mathematical approach, we can find an explanation in calculus. Zeno's essential premise that the arrow does not move at any particular point is not actually correct! To see why, we will delve a little into differential calculus. The following explanation is for those people who have either never learned calculus or forgot it, but it is always a fun one — so consider tagging along.

The *slope* of a line is how "steep" it is and is a measure of the rate of change. In the case of a graph of position vs time,

$$\text{Slope} = \frac{\text{Change in position}}{\text{Change in time}} = \text{Velocity.}$$

But that's for lines. What about curves? A position vs time curve which is linear implies a constant velocity, but we rarely see motion of constant velocity. How can we measure the slope of a curve in which velocity varies?

It turns out we can employ a concept very similar to the slope. But, we have to think outside the box. Consider taking two points: A and B, such that they are h units away from each other. Let's say we want to measure the slope at A.

We can get an estimate of the slope or velocity by using a *secant line*. At point A, or at the coordinate $(x_0, f(x_0))$, we have:

$$\text{Velocity} = \frac{f(x_0 + h) - f(x_0)}{h}.$$

To get a better and better approximation of the slope, we can decrease h. This is the core discovery of differential calculus: if we take the *limit* as h goes to 0, the slope of our secant line becomes the *instantaneous rate of change* of $f(x)$ at A. Mathematically, this can be expressed as:

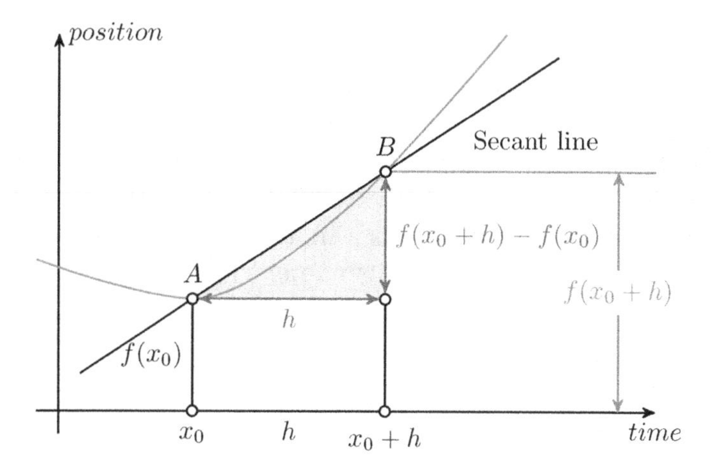

Figure 16: By taking the secant line between two points in a curve, we can estimate the tangent line whose slope will give us the rate of change.

$$f'(x_0) = \lim_{h \to 0} \frac{f(x_0 + h) - f(x_0)}{h}.$$

Here, the prime superscript on the f denotes that we are taking the *derivative*. The derivative of a function is another function that gives you the "slope" or instantaneous rate of change at each x_0. How does this apply to our resolution of the Arrow paradox?

It is our concept of the limit in the definition of the derivative. Notice that even though h approaches 0, which defines the "duration," there is still motion. Although time can be defined as a set of infinitesimal moments, there is infinitesimal movement in those moments. We do not need to invoke any rigorous theorems of calculus in this explanation, but we do need to refine our notions of what "motion" means. Motion, in this framework, can be seen as a functional relationship between time and space. By refining vague concepts, we come to a more solid logic that is less prone to contradictions like the one here.

We will see this time and time again throughout the book. Sometimes all it takes is some thinking about foundational ideas! This is the theme for much of mathematics, both for students and professional mathematicians.

The essence of the contradiction here is the issue of infinite processes. As mentioned previously, infinite processes were problematic for millennia. As a matter of fact, so was almost every thing with "infinite" in its name! Visit the infinity part of this book for some paradoxes regarding the infinite.

Achilles and the Tortoise

This paradox is arguably the most popular out of Zeno's paradoxes. It appears in mathematical literature historically more than any other — perhaps because adding characters to a story is more entertaining than studying the behavior of an inanimate arrow.

The argument to resolve this is rather simple and follows the same argument behind the geometric series resolution of the Dichotomy paradox. Suppose that Achilles the trackstar is 10 times faster than the slow tortoise such that he travels at 10 meters per second and the tortoise travels at 1 meter per second. Also assume that the tortoise has a 10 meter start. We can see that the distance at which Achilles will overtake tortoise is:

$$\begin{aligned} &10 + 1 + \frac{1}{10} + \frac{1}{100} + \cdots \\ &= 11 + .111\cdots \\ &= 11.111\cdots \\ &= 11\frac{1}{9} \end{aligned}$$

Combined with our insights about analyzing *instantaneous rates of change*, and knowing that differences in these rates of change is equivalent to differences in velocity, we can see how this paradox is resolved. Essentially, this paradox resolves itself through the insights we gained through resolving the first two.

3.3 LEGACY

Zeno's paradoxes, to some mathematicians, seem to be basic mathematical questions that can be easily resolved with calculus. And that is what appears to many students familiar with calculus. Even those students who do not have any notion of calculus can develop some basic conception of these paradoxes. It seems that Zeno's paradoxes can be wonderful thought-experiments to start looking into calculus.

However, many philosophers maintain their importance to this day, and believe that they capture the essence of many problems in metaphysics regarding the nature of space and time.

It is entirely possible that Zeno's paradoxes, due to their simplicity and universality, will always serve as a philosophical litmus test onto which people project their most fundamental philosophical and mathematical conceptions.

4

RUSSELL'S PARADOX: THE NECESSITY OF RULES

Imagine a game of soccer[1]. But now imagine it without any specific rules. "Intuition" would guide the refereeing of the game. No specific rules on what constitutes a foul, for example, would be used.

For a few games, this might pass just fine. But, eventually, this intuition would produce heated discussions on what the rules even are. And suddenly, a game that should have run perfectly ends in complete disagreement and argument.

The same goes for math: without a proper set of mathematical "rules," so called axioms, mathematicians would not get along too well. Eventually, contradictory opinions will emerge based on different understandings of what these axioms actually are. That is what mathematicians realized towards the end of the 19th century, and sought to change in the 20th century. A move towards establishing a foundation of mathematics was initiated.

How do these axioms erect mathematical structure? Let's think of science.

1 Or football as it is known internationally. A subtle "paradox" is behind this: American football is named "foot"-ball but the sport itself contains very little kicking.

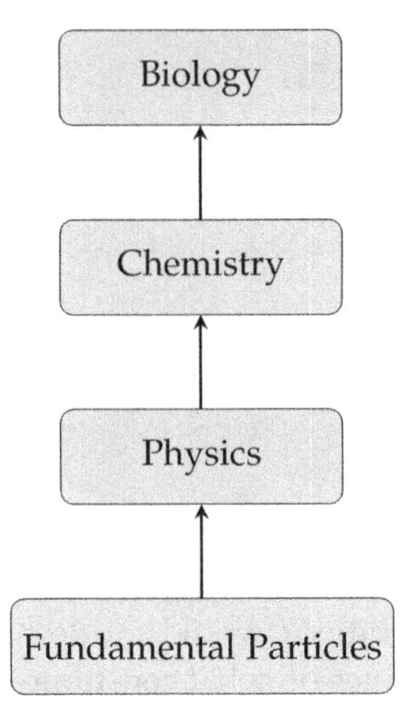

Figure 17: A "hierarchy" of the sciences.

In a scientific framework, the fundamental particles and their properties influence the whole that is science. In effect, they act as the "building blocks" in much the same way axioms would in a mathematical structure.

The end result is that the whole of science is influenced by these fundamental building blocks; as a building would be influenced by its foundation. Trust me, both of my parents are civil engineers! However, mathematical building blocks are not exactly easy to get right, and we will come across why later.

We will begin our discussion with a reformulation of Russell's paradox. Consider the following scenario:

> Paradoxville has only one barber. This barber shaves everyone who does not shave themselves, and only those individuals. Now, does that barber shave himself?

What's the paradox here? Well, the barber shaves all of those and only those who do not shave themselves. If he were to shave himself, then that would be a contradiction as the rule explicitly states he only shaves those who do not shave themselves. If he weren't to shave himself, then he again is breaking the rule as there is one person in town that doesn't shave themselves, but isn't shaved by the barber either: the barber himself. A contradiction!

Figure 18: Cartoon by Geoff Draper, taken from *Computation Engineering: Applied Automata Theory and Logic* by Dr. Ganesh Gopalakrishnan.

But the purpose of this book is not to discuss barbers — it is about math! So, what exactly is the mathematics behind all of this? It turns out that this scenario is simply an applied version of *Russell's paradox*. Russell's paradox is perhaps one of the most intriguing and influential paradoxes formulated in humanity's history. It was an attempt by the British logician, philosopher, and mathematician Bertrand Russell to show the inconsistencies of *naive set theory*, or the early framework of set theory formulated by Georg Cantor and Gottlob Frege. Before proceeding to the paradox, we will cover some of the history of set theory.

4.1 NAIVE SET THEORY

We now know why we need rules, so let's make some! Through the early work of Cantor and others, naive set theory was developed — quickly becoming a promising candidate for a theory for the "foundation of mathematics."

Cantor was the figurehead of the new move in mathematics to provide a foundational theory. Before the contributions of Cantor and Frege — which we will delve into in the next section — much of mathematics was disconnected. Analysis was very different from topology, which was very different from number theory. There was no common framework that fully united these branches of mathematics. Naive set theory was one such attempt, and although it satisfactorily laid the foundation of much of mathematics, it did not hold up to scrutiny.

In naive set theory, Cantor advocated the concept that any definable collection of elements is a set. For the sake of simplicity, these elements will be numbers, but they can be anything we desire them to be. Set theory is a very abstract theory and can be applied to all sorts of scenarios (as you may suppose, the concept of a collection of elements is just about as universal as you can get).

But naive set theory was defined in natural, informal language, and was a bit *too* flexible. As we recall from our rule-less soccer game, the more flexible rules are, the more trouble we are about to head in. Naive set theory stated that any set that can be defined exists. In most circumstances, this was completely fine. A set is one of the most elementary concepts we can imagine: it is simply a collection of elements. We can define a lot of sets very comfortably with the notation of set theory[2]:

2 See notation guide for more.

$$A = \{1, 2, 3, 4\}$$
$$B = \{x: 0 < x < 1\}$$
$$C = A \cup B$$
$$D = \{\text{All my followers on social media}\}$$

This general notion, as we will later see, is exactly what led to Russell's paradox — and many others in naive set theory.

4.2 RUSSELL AND HIS PARADOX

Bertrand Russell was a British polymath known for his impact on a variety of fields, from philosophy to mathematics to politics. He co-authored *Principia Mathematica*: a quintessential work of classical logic. Although not used widely today due to its complexity, it was behind many advances in mathematics.

Figure 19: Bertrand Russell (1872 - 1970)

Russell didn't like this free-roaming, informal framework. As a logician, he wasn't a fan of defining whole foundations of fields in informal language, which is what naive set theory did. Even though these early theories sufficed for everyday use of set theory in mathematical circles, they were considered too fragile by Russell. Therefore, Russell attempted to find a definable collection of elements that cannot exist as a set.

$*54{\cdot}43.\ \vdash :. \alpha, \beta \in 1 . \supset : \alpha \cap \beta = \Lambda . \equiv . \alpha \cup \beta \in 2$

Dem.

$\vdash . *54{\cdot}26 . \supset \vdash :. \alpha = \iota`x . \beta = \iota`y . \supset : \alpha \cup \beta \in 2 . \equiv . x \neq y .$

$[*51{\cdot}231] \qquad \equiv . \iota`x \cap \iota`y = \Lambda .$

$[*13{\cdot}12] \qquad \equiv . \alpha \cap \beta = \Lambda \qquad (1)$

$\vdash . (1) . *11{\cdot}11{\cdot}35 . \supset$

$\vdash :. (\exists x, y) . \alpha = \iota`x . \beta = \iota`y . \supset : \alpha \cup \beta \in 2 . \equiv . \alpha \cap \beta = \Lambda \qquad (2)$

$\vdash . (2) . *11{\cdot}54 . *52{\cdot}1 . \supset \vdash .$ Prop

From this proposition it will follow, when arithmetical addition has been defined, that $1 + 1 = 2$.

Figure 20: Excerpt from *Principia Mathematica*: proof that $1 + 1 = 2$. Russell and his co-author A. N. Whitehead go on to say that "The above proposition is occasionally useful." Who would have guessed?

Russell came up with the following:

Russell's Paradox

Let R be the set of all sets that are not members of themselves. If R is not a member of itself, then by definition it must contain itself. If it does contain itself, however, it contradicts its definition as only including sets that are not members of themselves. Contradiction spotted again! Symbolically, this can be represented as

$$\text{Let } R = \{x \mid x \notin x\}, \text{ then } R \in R \iff R \notin R$$

As you can see, the nature of this paradox is very similar to that of the Barber paradox. Russell's paradox is simply the abstract, original statement that motivated the Barber paradox as an applied analogue. Why do we care? From the principle of explosion in logic, any proposition, no matter how outlandish, could be proved from contradictions in a formal system. The presence of contradictions like these is a calamity for mathematics as it can topple the entire framework of mathematics, meddling with the ideals of mathematical truth and falsity. Like the false proof that $1 = 2$ in the first chapter, falsities can be manipulated in an infinite number of ways to produce an infinite number of outlandish conclusions.

4.3 THE TRAGIC STORY OF A GENIUS

Let's talk about the millennia of intellectual advances that led to Russell's paradox. There were a lot of attempts at foundational theories, so let's begin with proto-attempts at such theories. We are especially looking for the question: "What is mathematics?" We begin our story with Plato, a huge figure of philosophy and a prominent mathematics aficionado. Plato described mathematics as some sort of *mental reality*, in which numbers existed not in the world but in some alternate reality of concepts ("the world of forms").

Aristotle, his most famous student, disagreed. Aristotle thought that numbers themselves were not objects, but in fact properties of objects. So if I have five pencils in front of me, the pencils and the number of five weren't two separate objects, instead five is a property of the number of pencils on my table.

Fast forward 2000 years to Immanuel Kant, yet another huge mathematics aficionado. Kant disagreed with Plato's notion that mathematics is objective truth independent of human experience, and thought that to come up with any mathematics one would need intuition. Sure, mathematics described the world, but we also were *creating* math from our experience, thought Kant.

Then comes Frege, arguably the hero behind this whole story.

Gottlob Frege was a German mathematician, philosopher, and logician widely considered to be the father of analytic philosophy. He was very influential in the development of modern logic and the philosophy of mathematics. Although unrecognized during his lifetime, he is now celebrated for his invaluable contributions.

Figure 21: Gottlob Frege (1848 - 1925)

A reserved man and a quiet logician, Frege was overlooked in his own time. His ideas, however, proved fundamental and revolutionary. He disagreed with Aristotle's view of numbers as merely properties of objects. If they were, Frege thought, then only one number would belong to any arbitrary object — uninfluenced by subjective opinion. For example, take a deck of cards. Is one the number corresponding to the *one* deck of cards, or is 52 the number corresponding to the 52 cards in that deck? The ambiguity of this notion disproved Aristotle's theory to Frege.

Frege also disagreed with Kant and his intuition theory of mathematics, stating that arithmetic can be known from logic alone. With these prominent disagreements, he sought a new way of conceptualizing number and mathematics, based on logic. This initiated a whole movement in mathematics named *logicism*. Frege began by asking:

What is "number"? Define it without using any sort of circular definition (i.e. not including the word number, quantity, etc, in your definition).

Frege defined numbers using *concepts* and *extensions*. A concept is any idea you can think of, and an extension would be the set of all things belonging to the concept. A concept might be books, and the extension is the set of all books.

Numbers are then extensions of concepts, with each concept assigned a sort of "extension number." For example, all things made of three objects would have extension number three. Similar on the surface to previous mathematical philosophy, but yet very different in its details.

Frege took it as an axiom that each concept had a corresponding extension (even if it was empty, e.g. a square circle is a concept with an extension that is the empty set). This would become the *general comprehension principle*, which states that the number of concepts is identical to that of extensions. This sounds pretty reasonable, right? Not so fast.

Just as Frege published his work, Russell sent a letter to him stating that even though he agrees with him on quite a lot, he has just one difficulty: that which will later become his famous paradox. Russell succeeded in formulating an argument that would topple the Frege/Cantor regime, showing a grave contradiction.

As we saw previously, the language of set theory lends itself for the easy characterization of many different kinds of sets. But what about sets that *contain themselves*? More rigorously, what about sets that are in fact subsets[3] of themselves?

Normally, sets are not members of themselves. For example, think of the set of all living people. That set does not contain the set of all living people as a member. Call a set like this a *normal* set. Can you, however, come up with a definition of a set that is in fact a member of itself — an abnormal set?

3 Subsets are simply a set of elements that are contained within the "parent set." For example, $\{1,7\}$ is a subset of $\{1,2,3,5,7\}$. More on this and set theory in general can be found towards the end of the book.

As it turns out, in naive set theory, you can define such a set. This was naive set theory's death sentence — its whole structure will now be easily toppled.

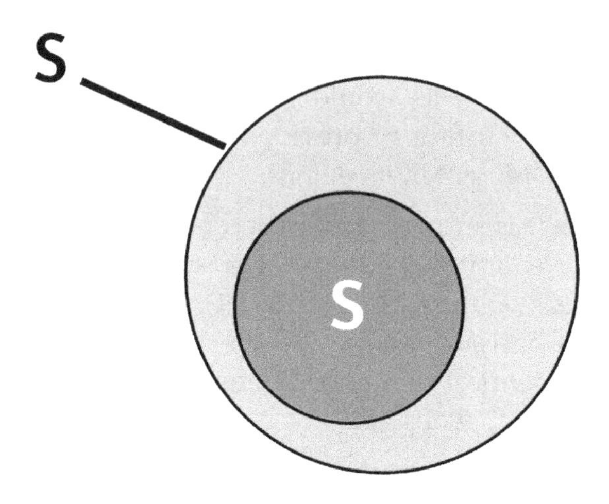

Figure 22: Set S contains itself. Venn diagrams are a popular tool in naive set theory for illustrating concepts effectively and intuitively.

It is in fact pretty easy to define such a set using naive set theory's formalism:

$$S = \{1, 2, 3, S\}$$

Now, is the *set of all sets that do not contain themselves*, a subset of itself? As we reasoned in the last section, if that set is not a member of itself, then by definition it must be a member of itself. If it does contain itself, then it contradicts its own definition.

Frege crumbled after this. It seemed like his whole life's work was made futile by a single paradox. Frege was in fact so crushed by this he had a mental breakdown which eventually took him to the hospital.

However, Frege later took this incredibly well. He hurried to make an appendix mentioning Russell's note, describing it as well as the fact that its implications are not quite clear as of the moment. Eventually, Frege would have to give up many of his ideas on the foundations of mathematics. In a bitter irony, Frege had actually written earlier about this possibility, stating:

> *We must really face the possibility that we may still in the end encounter a contradiction which brings the whole edifice down.*

In writing about Frege's response to the paradox, Russell wrote:

> As I think about acts of integrity and grace, I realise that there is nothing in my knowledge to compare with Frege's dedication to truth. His entire life's work was on the verge of completion, much of his work had been ignored to the benefit of men infinitely less capable, his second volume was about to be published, and upon finding that his fundamental assumption was in error, he responded with intellectual pleasure clearly submerging any feelings of personal disappointment. It was almost superhuman and a telling indication of that of which men are capable if their dedication is to creative work and knowledge instead of cruder efforts to dominate and be known.

4.4 RESOLUTION

About seven years later, two ways of avoiding the paradox were proposed: Russell's type theory and the Zermelo set theory. While Russell advocated for an alteration of the logical language and framework itself, Zermelo simply went beyond Frege's flexible axioms. Zermelo would form the first *axiomatic* set theory, which is a set theory based in formal logic. Zermelo set theory evolved into the canonical Zermelo–Fraenkel set theory (ZFC if we also add in the axiom of choice, which we will see come up in the next chapter) which forms the basis of mathematics today. In a happy ending, a lot of Frege's original ideas are incorporated into ZFC, so his research was not fruitless at all.

But the story is not over. ZFC is not free from criticism, and some newer theories are proving to be more lucrative candidates for a "foundation of mathematics" day by day. One such theory is category theory, which abstractly represents mathematical concepts through "categories." Another is homotopy type theory, which revives Russell's work on type theories, albeit in a different form.

4.5 GRELLING'S PARADOX

Other than the barber paradox, there are a few other paradoxes that are very similar, even equivalent, to Russell's paradox. We will discuss only one here: Grelling's paradox.

Grelling's paradox mirrors Russell's in many ways, but it never gained the same popularity. Why? Although pointing at a very fundamental concept, it was seen only through a semantic and linguistic sense rather than in any mathematical framework.

Consider the two words *autological* and *heterological*, both of which describe adjectives. As a reminder, an adjective is a description of a noun. For example, take the sentence "We are examining a weird paradox." In that sentence, the adjective "weird" is a descriptor of the noun "paradox." But does weird describe itself? Is the adjective "weird," weird?

Here we see where the lengthy words above come into play. An autological adjective describes itself. For example, the adjective "English" is English and describes itself. A heterological adjective, however, does not describe itself. For example, the adjective "hyphenated" is not itself hyphenated, so it does not describe itself.

Grelling's paradox can be translated as the following: identify each adjective as corresponding with the set of all objects it describes. This is very much similar to Frege's theory of concepts and extensions. For example, the adjective "green" would correspond with the set of all green objects. Consider the adjective "English," which corresponds with the set of all objects that can be described as English — i.e. all the words in the English dictionary and their possible permutations. English's corresponding set in fact contains "English." Thus, an autological adjective can be understood as an adjective whose corresponding set contains the adjective itself. This is because it describes itself.

The question of whether the adjective "heterological" is itself a heterological word is in fact very similar to Russell's paradox. If it is not, then "heterological" must be autological. This leads to a contradiction, however, as "heterological" does not describe itself here (we just reasoned it is autological, or the opposite, so it must be heterological). If it is in fact heterological, it would also lead to a contradiction. Now, the word "heterological" describes itself — therefore it is autological.

Yes, that was quite difficult to navigate through. Nonetheless, it very much embodies the same concepts that led to Russell's paradox. While Grelling's paradox concerns words and Russell's paradox concerns sets — they both aim at the same fundamental idea.

5

GÖDEL'S INCOMPLETENESS THEOREMS

We have spent a good deal of time discussing axioms and the foundations of mathematics, but we have not paused to think if mathematics itself is limited in any way. No, not limited because of our lack of ingenuity or mental prowess, but rather limited because of some inherent characteristic. Does mathematics, this seemingly universal language, also have its limits? It is of course possible that this question did not occur to anyone throughout history due to mathematics' prowess! Just about everything — from linguistics to engineering — can be described by mathematics. But could there be statements that our mathematical structure could never prove, even though they are true?

In this chapter, we will be talking meta-everything. We will be looking at the inherent limits of mathematics and the ingenious insights of one person: Kurt Gödel. Gödel determined that formal systems of mathematics could not be complete, i.e. able to prove everything that is true, and could not prove their own consistency. But, before we delve into his ideas in more depth, we ought to cover some background.

5.1 MATH WARS

As we alluded to previously, the late 19th and early 20th century saw the rise of the *foundational crisis* in mathematics. The rise of several paradoxes such as Russell's paradox motivated a massive search for a foundational theory of mathematics.

The only subfield of mathematics that truly had an axiomatic basis was geometry, a field established more than 2000 years ago by Euclid in his *Elements*. Euclid's *Elements* presented the crucial axiomatic method which was considered as the best model of what mathematics should look like. Other mathematicians, after mathematics became more rigorous, began to question whether other subfields of mathematics could indeed be just like geometry. They wanted that axiomatic structure for all mathematics — and they wanted it bad.

But, as with any mathematical issue with this much at stake, stark disagreement emerged. Several schools of thought existed, and perhaps one of the most well-known "math wars" emerged. It was fierce, to say the least.

The leading school was the formalist school, which was headed by David Hilbert. This group was in fierce opposition with the intuitionists, led by L. E. J. Brouwer. The intuitionists regarded the formalist program as a meaningless boatload of symbols. This controversy culminated in Hilbert successfully removing Brouwer from the editorial board of Mathematische Annalen, the leading mathematical journal of the time. In total, there were three prominent schools of thought at war with each other: the logicists — which we have already discussed briefly — , the formalists, and the intuitionists. We'll take a brief look into their beliefs and the history behind their movement in the next three subsections.

5.1.1 *Logicism*

We have already encountered logicism in the previous chapter. It is perhaps easiest to define — it is simply the belief that mathematics is either an extension of logic, or can be entirely reduced or modelled on logic. Russell and Whitehead in their *Principia Mathematica* perhaps best embodied its spirit. The logicians were quite powerful and influential, but they had some pretty good competition: the formalists.

5.1.2 *Formalism*

In 1900, at the outset of the 20th century, David Hilbert delivered a famous speech at the International Congress of Mathematicians: the biggest conference dedicated to mathematics. He had a passionate vision for this century and identified a total of 23 unsolved problems that he hoped to see solved. This is the very well-known "Hilbert's list."

Although aware of the problems and paradoxes plaguing naive set theory, he had no idea how difficult his second problem — constructing and proving a set of axioms to be consistent — will prove to be[1].

Hilbert, who we will get to know better in the next part of the book, was one of the most influential mathematicians of the late 19th and early 20th centuries. As we saw previously in the case of Russell's paradox, that very same time period was when mathematicians realized that they needed a proper foundation for mathematics. With paradoxes coming in left and right, Hilbert sought to ground all existing theories to a finite set of axioms, as well as prove that these axioms would make for a consistent mathematics. This was the famous *Hilbert's plan*.

Hilbert's goal was to prove the consistency of very complex systems, such as real analysis, using simple systems. His

1 Of those problems that are considered definite enough to make a judgement on whether they were resolved or not, nine have resolutions that are accepted widely. Seven problems have resolutions with partial acceptance (including the second), and four are still unresolved. A close parallel to Hilbert's 23 problems are the 7 Millennium Prize problems, which were also announced at the beginning of a new century, but have a lucrative 1 million dollar prize attached to the first correct resolution. I should warn you, however, that this is perhaps one of the hardest ways to become a millionaire.

eventual goal was to reduce all mathematics to basic arithmetic[2]. He believed that, if set up correctly, this system would be both consistent — avoiding contradictions — and complete — being able to prove all that is true.

Hilbert even admitted to formalism being some sort of "formula game" in a response addressing Brouwer, stating:

> *And to what extent has the formula game thus made possible been successful? This formula game enables us to express the entire thought-content of the science of mathematics in a uniform manner and develop it in such a way that, at the same time, the interconnections between the individual propositions and facts become clear... The formula game that Brouwer so deprecates has, besides its mathematical value, an important general philosophical significance. For this formula game is carried out according to certain definite rules, in which the technique of our thinking is expressed. These rules form a closed system that can be discovered and definitively stated.*

This seems satisfactory, but nonetheless poses significant unanswered questions. Why this particular set of axioms? Why this specific logic? The fellow German mathematician Hermann Weyl would ask similar questions about Hilbert's approach:

> *What "truth" or objectivity can be ascribed to this theoretic construction of the world, which presses far beyond the given, is a profound philosophical problem. It is closely connected with the further question: what impels us to take*

2 In this book, arithmetic in the context of the foundations of mathematics shall refer to the full rich universe of the properties of natural numbers and their relationships with each other, not merely the mechanical processes associated with arithmetic. In other words, arithmetic is occasionally used as a synonym of *number theory*. For example, despite Fermat's last theorem being a statement about basic arithmetical relationships, the methods employed in its proof are much more sophisticated than arithmetic operations. Yet, these methods are based on basic operations of arithmetic.

> *as a basis precisely the particular axiom system developed by Hilbert? Consistency is indeed a necessary but not a sufficient condition. For the time being we probably cannot answer this question ...*

5.1.3 *Intuitionism*

The intuitionists take an entirely different approach — to the intuitionist, mathematics is an invention of the mind. Without humans, numbers would not exist. In this framework, math is not considered an analytic field where deep truths about objective reality are revealed, but instead is the application of methods deemed "useful" to realize complex mental constructs regardless of their independent existence in objective reality. Yes, that was heavy. But essentially — to the intuitionists mathematics does not represent any truths about objective reality and is an intrinsically subjective activity (hence the word "intuition").

The intuitionists were very controversial — and for good reason. They objected to the principle of excluded middle, which is one of the most fundamental laws of logic. Essentially, that principle states:

> For any proposition p, either p is true or $\neg p$ (its negation) is true.

Hilbert waged war against this notion — and the intuitionists at large. He expressed his thoughts in the foundational text *Grundlagen der Mathematik,* stating:

> *Taking the principle of excluded middle from the mathematician would be the same, say, as proscribing the telescope to the astronomer or to the boxer the use of his fists.*

5.2 GÖDEL'S SURPRISE

Then came Gödel.

Kurt Gödel was an Austrian logician, mathematician, and philosopher considered to be one of the greatest logicians in history.
Gödel's incompleteness theorems, along with other contributions, completely transformed mathematical logic. As one of Einstein's close friends, he also took interest in general relativity and proved the possibility of time-travel.

Figure 23: Kurt Gödel (1906 - 1978)

Yes, you read that right, he did prove the possibility of time-travel in the framework of general relativity[3].

As a doctoral student at the University of Vienna, Gödel was hesitant to accept the conclusions of Russell and Whitehead in their *Principia Mathematica*. That book was the pinnacle of the movement towards the formalization of the foundations of mathematics — especially that of the logicists. Similar to what we saw in the previous chapter, this was a move towards formalization and an axiomatic mathematics rather than a naive one based on loosely-defined words. *Principia Mathematica*, in effect, tried to create an all-encompassing mathematical language,

3 Gödel showed that, under certain conditions, a universe with "closed world lines" is possible. This would be a universe in which time has a circular structure and objects at some point seamlessly return to their own past. Thus, Gödel essentially proved the mathematical possibility of time travel within the framework of general relativity.

with symbols being its diction and the system of rules, or axioms, governing them the grammar. Gödel doubted this.

In studying *Principia Mathematica*'s uninviting complexity, Gödel came to a shocking discovery. He discovered that he could reformulate the entire 700 or so pages of *Principia Mathematica* into numbers. He reimagined the rigorous logical arguments in this crucial text from being an array of symbols to being based entirely on numbers. He then would find out that the system *Principia Mathematica* tried to describe — arithmetic — is in fact what it is based on. *Principia Mathematica* was using arithmetic to describe arithmetic.

This will sound a little difficult to comprehend at first, but bear with me. In looking at *Principia Mathematica* in a different way using numbers, Gödel saw *Principia Mathematica* as its own system. He saw that the patterned arguments present in the countless pages frenzied with symbols *themselves* say things about each other and possibly even say things about themselves. Ironically, Russell's magnum opus proved to be one of the biggest "self-references." Russell's metamathematics used numbers to prove general truths about numbers. Yes, that was confusing. Let's delve into a little background to demystify those statements.

5.2.1 *Background*

The Liar Paradox

In gaining that insight about *Principia Mathematica*, Gödel was reminded of an early logical paradox from the Greeks: the *Liar Paradox*. The liar paradox dates back to around 500 B.C. in a poem by Greek philosopher and poet Epimenides. The famous claim was:

> All Cretans[a] are liars.
>
> *a* Cretans are inhabitants of the Greek island Crete.

But what's paradoxical about this statement? The thing is — Epimenides was himself a Cretan! Historically, Epimenides probably did not intend this to be a paradox, but rather a poetic take at Cretans for denying the immortality of the Greek god Zeus. But, strictly looking at this statement and assuming a "liar" is one that always speaks falsehoods and a "truth-teller" is one that always speaks truths, we can see the paradox emerging. English academic Thomas Fowler described the statement's absurdity in his book *The Elements of Deductive Logic*:

> *Epimenides the Cretan says, 'that all the Cretans are liars,' but Epimenides is himself a Cretan; therefore he is himself a liar. But if he is a liar, what he says is untrue, and consequently, the Cretans are veracious; but Epimenides is a Cretan, and therefore what he says is true; saying the Cretans are liars, Epimenides is himself a liar, and what he says is untrue. Thus we may go on alternately proving that Epimenides and the Cretans are truthful and untruthful.*

This paradox is not some ancient poetry relic — it pervades our everyday life without us noticing! As a kid, which I guess I still technically am as I am only 17, there was always that rebellious friend with little to no regard for rules. This phrase might come up with them:

All rules are made to be broken.

Or

All rules have exceptions.

Upon further analysis, we see that the statements above always seem to end in contradiction. If all rules have exceptions or are made to be broken, then the above rules must have exceptions or are made to be broken too.

Let's go back to our original example. If the statement is true, then by definition it is false because all Cretans do is lie! This is a contradiction. However, we can say that it is false without a contradiction. Negating the statement "All Cretans are liars" results in the statement "Not all Cretans are liars." Someone who is not a liar does not necessarily have to be a truth-teller — a person that makes both true and false statements is neither a truth-teller nor a liar. Therefore, we can resolve this paradox if there is at least one Cretan who is a non-liar.

But, what if there is only one Cretan? What if we *strengthened* our statement to be the general:

This sentence is false.

Is this statement true or false? This is the crucial liar paradox Gödel used to build up his now-famous insights regarding mathematical logic.

Figure 24: Cartoon obtained from www.sketchplanations.com

If we try to resolve this paradox, we encounter an endless cycle. If the statement is indeed false, then it must be true. But if it is true, then by the statement's declaration it is false. We can keep going on with this forever. Trying to assign a binary True/False value to this statement almost seems impossible.

There are countless statements like these that defy logic. They are named *self-referential statements*. We saw a similar form of this *self-reference* in our previous chapter regarding Russell's paradox. The same endless cycle that we encounter in trying to determine the validity of a statement is embodied in trying to define a set which is self-referential. Gödel used these statements to build up his theories regarding logic and the foundations of mathematics.

To illustrate what is problematic about self-reference, consider this. Let's say you walk into a room filled with top-secret documents. In that room, there is a sign saying "Do not read anything in this room." In everyday experience, this would still get its message across. However, one might not realize that they already read something in that room, markedly the sign. But if you do not read the sign, then you will not be aware of the command. That is unless some other entity told you this word of caution — a human for example. A human saying "Do not read anything in this room" is perfectly valid logically and contains no subtleties that render it problematic. We now see that the issue is not one of what is said, but how it is said. Self-reference is the problem.

The Dilemma of Consistency

The Greeks left us three important problems that remained unsolved for over 2000 years. They were:

- Trisect any angle with a compass and a straight-edge.
- Construct a cube with a volume twice the volume of any given cube.

- "Square the circle," or construct a square that is equal in area to a given circle.

Now, these problems are all elementary. But it took millennia for mathematicians to solve them. This was all a consequence of the intense expansion of the mathematical adventure in the 19th century. This, of course, stimulated much more mathematics as the cycle of math goes. The field of real analysis, as well as many other fields, emerged.

But there was an even bigger revelation that was a result of the resolution of some old mathematics — *Euclid's fifth postulate was false*. As aforementioned, Euclid gave the first axiomatized field of mathematics, namely geometry. In the beginning of his *Elements*, Euclid gave a set of five axioms that he would base the entirety of later proofs on. They can be essentially understood as the following:

- A straight line segment can be drawn to join any two points.

- Any straight line segment can be extended indefinitely in a straight line.

- Given any straight line segment, a circle can be drawn having the segment as its radius and one endpoint as center.

- All right angles are congruent.

The first four are pretty reasonable, right? The fifth one, however, troubled thinkers for millennia:

- If a line segment intersects two straight lines forming two interior angles on the same side that sum to less than 180°, then the two lines, if extended indefinitely, meet on that side on which the angles sum to less than 180°.

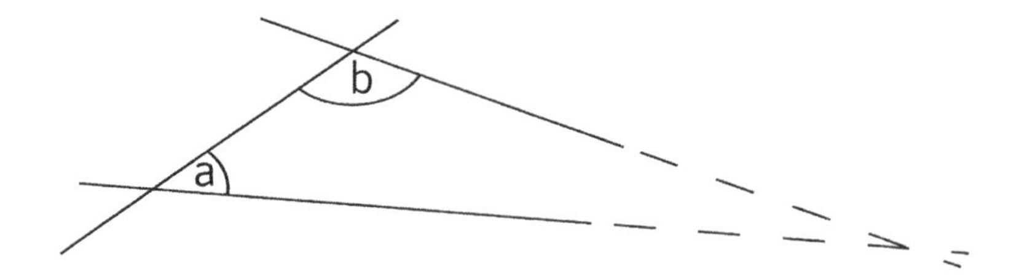

Figure 25: Illustration of the "parallel postulate." Notice that since $a + b < 180°$, the lines intersect on their side.

Still makes sense, right? But, the fifth postulate just does not hold the same ring to it when compared to the others. It does not seem so self-evident. It was widely believed that this postulate could be proved from the first four. But nothing came out. Hence, mathematicians sought for thousands of years to prove this fifth postulate from the first four.

Hundreds of mathematicians laid their hands on this problem. This problem was especially of interest to mathematicians in the Islamic Golden Age, including the likes of Ibn al-Haytham, Omar Khayyam, and Nasir al-Din al-Tusi. Even then, no valid proof was found. Perhaps a proof cannot be found! But, let's not take that view yet — we have to stay optimistic. As Thomas Edison put it when asked about his failures (paraphrased):

> *I have not failed 1000 times. I have successfully found 1000 ways that do not work.*

Nonetheless, through the work of Riemann, Gauss, Bolyai, and many others in the 19th century, it was demonstrated that it is in fact impossible to deduce the parallel axiom from the first four axioms. It simply did not follow.

This was revolutionary for two reasons. First, it demonstrated that it can be *proved* that something is unprovable, all within the same system of mathematics. As we will see later, proving that a statement is unprovable is the heart of Gödel's argument behind his incompleteness theorems. Second, it showed that we have a *choice* in what axioms we dictate our mathematics to follow. Euclid was definitely not the last word on geometry, and the notion that the axioms of geometry and mathematics at large can be reduced to whatever seems "self-evident" seemed like a deficient approach. Ironically, mathematicians — who prided themselves on creativity — realized that they were thinking inside a box for the entirety of more than two millennia. This box was Euclid's axioms.

This realization ushered a revolution in geometry and the expansion of its scope. Suddenly, hyperbolic and elliptic geometry were the center of focus. And these geometries ended up being crucial to later developments of science, markedly in Einstein's theory of general relativity. Alongside establishing axiomatic systems in geometry, this spirit permeated the mathematical community. The foundational crisis' origins can be traced here.

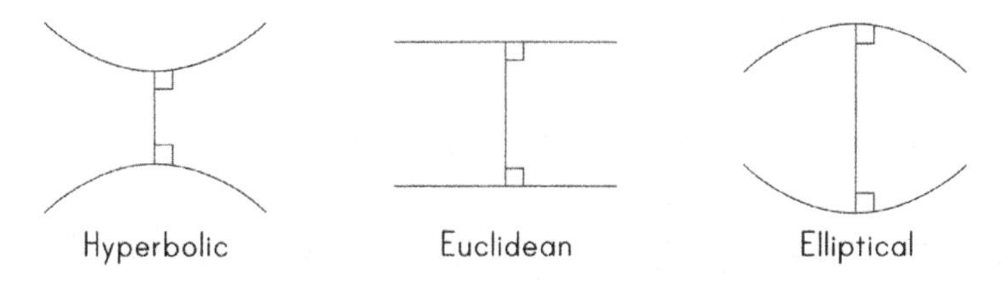

Figure 26: The difference between 1) Hyperbolic geometry, 2) Euclidean, or flat, geometry, and 3) Elliptical geometry.

Mathematics took a sharp turn towards the abstract here. Mathematicians saw that mathematics is not merely the science of quantity, and mathematical conclusions do not depend upon any meaning associated with the terms and symbols of math. Furthermore, mathematicians understood that the postulates of mathematics are not inherently about anything in our world, and concepts such as quantity, space, etc, play no role in "doing mathematics," or deriving theorems. Mathematics became much more formal, mirroring logic and philosophy. Mathematics proper is now to be seen not as determining the validity of axioms or even conclusions, but determining if a conclusion is a *logical deduction* from a set of premises or axioms.

While the scientist uses mathematics to understand patterns in the physical world, the modern mathematician's primary role turned to being a theorem-generator. Paul Erdős, one of the most prolific mathematicians of the 20th century, perhaps best summarizes this sentiment:

> *A mathematician is a device for turning coffee into theorems.*

This opened up a plethora of new fields of mathematics. Considerable variety emerged in mathematics, even when that variety did not "make sense." Mathematicians recognized that intuition is but an ever-changing byproduct of what we are exposed to, so what seems counterintuitive now may not be in a century. Also, we know from the first two chapters that intuition cannot guide the search of mathematical truth, and perhaps truth in general — it simply leads to too many falsehoods in our intricate world.

But wait — now that mathematicians concern themselves only with generating theorems, whose truth is contingent on nothing but the starting assumptions, how could we know that this system that we are working under is consistent? If we are to make mathematics based in formal logic, we must obey one of its most important principles: the principle of non-contradiction (PNC). If we were to have any hope of not plunging into mathematical chaos, our system must be constructed such that we cannot deduce two mutually contradicting statements from it. We cannot have a reliable mathematics that violates the PNC.

When thinking about familiar objects such as points and lines, this does not seem like a big problem. But as we saw in the chapter on Russell's paradox, even intuitive truths about familiar objects can lead to contradiction. Sets are very familiar objects, and Frege's ideas were nothing outlandish, but they led to contradictions.

Let's remind ourselves that mathematics moved beyond the familiar in this era. Euclidean geometry's axioms may be easy to formulate, but when dealing with, say elliptic geometry, the problem of formulating a consistent set of axioms becomes much more difficult. Take this presumption from elliptic geometry: through a given point outside a line, *no* parallel to it may be drawn. That is definitely not what we see in our everyday experiences with geometry, so how does one go about proving the consistency of the set of axioms for elliptical geometry? More importantly, how can we prove the consistency of *Principia Mathematica*?

5.2.2 *The Toppling of a Regime*

In 1931, Gödel would publish the revolutionary paper "On Formally Undecidable Propositions of Principia Mathematica and Other Systems." However, the paper gained little traction at first. Gödel was, in a sense, way ahead of his time. The methods employed in his proof were so novel that only a select group of specialists at his time could follow his argument.

Now, as you may already suspect, understanding the full mathematical argument Gödel posed would still be difficult for any non-specialist. Nonetheless, the structure and essence of *Gödel's proof* is one that can be understood by virtually all readers with a basic knowledge of mathematics.

Gödel realized that he could write a formula that is unprovable by the rules of *Principia Mathematica* — not because he or other mathematicians at the time lacked ingenuity, but because of the inherent limits of Russell and Whitehead's system. This dismantled the entire goal of *Principia Mathematica*. Russell and Whitehead made it one of their primary goals to eliminate vicious circularity, but somehow it snuck through the backdoor to their system. The vicious circle principle can be understood as avoiding *circular logic*, e.g. in the previous chapter when we tried to define number without using the word number or any of its synonyms. More precisely, the principle states that no object or property may be introduced by a definition that depends on that object or property itself. It is this vicious circle that Russell and Whitehead saw in many of the paradoxes of their time — specifically those self-referential ones.

But, is this statement that Gödel came up with just another paradox inherent to *Principia Mathematica*? We can just come up with a new system then, right? Gödel set out to prove that his observation was much more than a criticism of the specific system erected by Russell and Whitehead but a universal observation on all such systems. He showed that this Gödelian formula was a true statement that cannot be proven using the *Principia Mathematica* system as well as *any* other system that can be mirrored in *Principia Mathematica*. Gödel did not only dismantle the goals of *Principia Mathematica*, but dismantled the goals of the entire movement it characterized. Single-handedly at 25 years old, Gödel proved what could have never been imagined before. The core of his demonstration is his realization that the systems of logic erected by Russell and Whitehead can be translated to numbers — this is where Gödel's numbering comes into play.

Gödel's Numbering

A Gödel numbering is simply a particular encoding which assigns numbers to logical/mathematical notation. Through some mechanism, another natural number can be assigned to a sequence of symbols. The idea behind this is if we can encode logic in arithmetic, we can manipulate logic arithmetically.

A simple analogy would be the encoding of characters and symbols in ASCII[4]. For example, the string "math" converts into 109 97 116 104 in ASCII.

4 Abbreviation for "American Standard Code for Information Interchange," a character encoding standard for electronic communication.

Symbol	$\neg$	$\vee$	$\forall$	(	)	...
Number	5	7	9	11	13	...

Table 2: Gödel's original encoding of some mathematical symbols.

Similar to ASCII, Gödel aimed to make the Gödel number of each statement to be unique and meaningful. Although these numbers will inevitably be long, that was not the issue — the issue is how we can construct such numbers to be meaningful. The most important goal behind developing this numbering system is to be able to do some number theory to deconstruct the validity of statements. Gödel saw a frenzy of symbols and sought to turn it back into the original language of mathematics: number.

This is a revolutionary insight. But, does it work? How complicated would these truth-determining number properties be? Most importantly, how can we make these numbers unique?

Gödel sought his building blocks to be identical to those of arithmetic: prime numbers. Prime numbers were perfect for Gödel's purpose due to their unique properties. Markedly, the fundamental theorem of arithmetic guarantees that each number has a unique prime factorization. Therefore, it is always possible to decipher from a certain number the notation behind its corresponding statement. There is much more to this notion of Gödel numbering, especially in the case of studying whole proofs, but we shall leave it at here since it becomes significantly more complicated. The key insight here is that *Principia Mathematica* can be interpreted as using numbers to talk about numbers, yet still claims its basis is not arithmetic.

Proof Summary

For the first incompleteness theorem, one can summarize Gödel's reasoning in three steps:

- Statements in the *Principia Mathematica* system can be represented by natural numbers, whose properties would be equivalent to determining whether their corresponding numbers have certain properties.

- In that system, or in general the formal system which can be mirrored in *Principia Mathematica*, it is possible to construct a number such that its corresponding statement is self-referential and essentially says "This statement is unprovable" using the technique of diagonalization, which we will look at later.

- This shows that the system cannot be consistent, therefore the original assumption that it is consistent is false.

The second incompleteness theorem can then be obtained by formalizing the proof of the first incompleteness theorem within the formal system itself. This process is quite complicated, but the conclusion of the theorem is not that difficult to understand.

The Statement that Ended it All

> This proposition is unprovable using the language and grammar of *Principia Mathematica.*

If this statement was provable, then it would be false. This means that Russell and Whitehead came up with an inconsistent system! If this statement was indeed unprovable, then it is true, and we just demonstrated that Russell and Whitehead's system is not complete. Checkmate.

This came as a shock to the mathematical world, and to just about everyone who ever studied mathematics. How could mathematics be either incomplete or inconsistent? Many initially disagreed, or tried to reformulate, Gödel's conclusions.

Ultimately, Gödel was right — not just about Russell and Whitehead's system, but about any system resembling the logicist dream. This gave rise to his two incompleteness theorems:

First Incompleteness Theorem

Any consistent formal system F within which a certain amount of elementary arithmetic can be carried out is incomplete; i.e., there are statements of the language of F which can neither be proved nor disproved in F.

Second Incompleteness Theorem

Assume F is a consistent formalized system which contains elementary arithmetic. Then F cannot prove its own consistency.

5.3 HILBERT'S DREAM: POST-GÖDEL ERA

Gödel's paper was truly the start of a new epoch in how we view mathematics. The once inspiring dreams of Hilbert and Russell to encapsulate mathematics into a system that is both consistent and complete now seem completely quixotic. Gödel forces us to reconsider the most fundamental aspects of mathematics and beyond.

After Gödel, what is mathematics anyways? What even is mathematical truth? What is truth in general? Gödel did not confine his proof to some exotic field of mathematics, he based it on its origin: arithmetic. Just like a building with a shaky base, if arithmetic is not consistent then there is no way for mathematics as a whole to be so.

Gödel's conclusions still allow us to discover new things regarding mathematics to this day. Clearly, his work is of lasting importance. It is important to recognize that Gödel did not per se discredit Hilbert's dream as a fiction, but rather proved the impossibility of constructing a proof of consistency for a formal system such as *Principia Mathematica* that *can be mirrored inside this work*. Gödel's work then does not strictly prohibit Hilbert's dream, but renders it unlikely. No one today seems to have a good idea of what a proof of consistency that could not be mirrored in *Principia Mathematica* would look like.

You might be wondering why we can't simply add whatever unprovable, but true, statement as one of the axioms of our formal system. Why can't we just "patch" our system?

The problem is if we add whatever statement that is, then we can simply make the statement:

> This proposition is unprovable within the new formal system.

Paradoxically, for every patch we fix, we create a new one!

You might wonder if Gödel's conclusions actually translate into anything more "concrete," such as a mathematical theorem. It turns out, there are statements that, in certain systems, are not provable yet true. However, when placed into a larger system with more axioms, they can be proved. The first example of such a statement comes from **Ramsey theory**, a branch of mathematics named after the mathematician and philosopher Frank P. Ramsey. Ramsey theory studies the relationship between order and disorder, and is considered a subset of combinatorics[5].

In 1977, the *Paris–Harrington principle* in Ramsey theory was proved to be undecidable under the axioms of Peano arithmetic but decidable in a larger framework of second-order arithmetic.

Beyond that theorem, there exist many current problems that mathematicians believe are true but unprovable in the standard axiomization of mathematics, i.e. ZFC set theory.

Gödel's work finds its way to many other related fields from physics to computing to philosophy. Gödel's foundational work

5 Combinatorics is an area of mathematics primarily concerned with counting. The most basic ideas from that field include permutations and combinations.

should not be misconstrued as a cause for pessimism, but rather should be celebrated for it shows the power of human reason to zoom out and critically evaluate systems of reasoning that have guided our intellectual development for millennia.

Part III

INFINITY

The Infinite! No other question has ever moved so profoundly the spirit of man; no other idea has so fruitfully stimulated his intellect; yet no other concept stands in greater need of clarification than that of the infinite.

— David Hilbert

Figure 27: The English mathematician John Wallis is credited with the introduction of the infinity symbol, ∞, to mathematics. Wallis never explained why, but it is conjectured that the inspiration for this symbol came from the ancient Roman numeral representation of 1000.

This story, like so many others in mathematics, originates with the Greek mathematicians of the classical era. Nonetheless, the Greeks endeavoured to avoid the infinite by all means. A surprising example is given by Euclid in his proof of the infinitude of prime numbers. The proof is short and lovely, so a brief summary of Euclid's argument will be given below.

There is no last prime

Proof. Assume that there is a finite set of prime numbers given by

$$\mathbb{P} = \{p_1, p_2, \cdots, p_n\}$$

Now, define

$$N = 1 + \prod_{k=1}^{n} p_k = 1 + p_1 \times p_2 \cdots \times p_n$$

Clearly, N is larger than our largest prime, p_n. Thus, it must be divisible by some p_k in our set $\mathbb{P}$. However, dividing N by any number from the finite set $\mathbb{P}$ gives a remainder of one, implying that this number is prime, and that there is a larger prime than p_n. This is a contradiction. Therefore, there are infinitely many primes.

Interestingly, Euclid avoided saying the word "infinite." Rather, he stated that he proved that there is no *last* prime. The subtleties behind the works of many mathematicians in this era show us that infinity was avoided, and perhaps for good reason. It did not seem to operate under the standard framework of number.

Various philosophical questions limited the scope of human understanding and sometimes resulted in the outright dismissal of the infinite. If the world is finite, which was assumed at the time, how could one make sense of infinity? Drawing back to our discussion of what "number" is in the chapter on Russell's paradox, if numbers were to be considered abstractions or properties of objects like Aristotle thought of them, what object would infinity be abstracted from?

Zeno's paradoxes could have easily been put into this part, but I have felt that they go beyond the scope of discussing infinity and its properties. Nonetheless, they were one of the first paradoxes of their kind. And they continue to show us the absurdities of the infinite to this day.

Infinity as a concept remained far beyond our reach for millennia. We can attempt to think about big numbers: the number of grains of sand on all earth's beaches, the number of fish in the ocean, or the number of stars in the universe. Those numbers are big, *huge* even, but they are finite. Infinity is detached from human experience, so the difficulties humanity had, and still has, in understanding it is a natural consequence.

Perhaps the most common misconception about infinity is that it is a *number* in the traditional sense. It is not! That is why our everyday intuition on how to handle numbers, namely the arithmetic skills we are taught in our elementary schooling, simply does not apply to the concept of the infinite. You can't do arithmetic with infinity, at least not in the traditional sense of the word.

Rather, infinity is simply a concept. Gauss, a prominent mathematician in the era of so many mathematical discoveries about infinity, was a fervent advocate for dealing with infinity as what it truly is: a mere concept.

> *I protest above all against the use of an infinite quantity as a completed one, which in mathematics is never allowed? The Infinite is only a manner of speaking.*

Nonetheless, infinities can be classified as a *certain type* of number. We will later see that infinities can be described as *transfinite* numbers, but those "numbers" obey properties that are much different from the laws of arithmetic that govern the real numbers we are used to.

Even though infinity is strictly outside of the realm of sensory perception, it is entirely within the realm of understanding. Through carefully constructed arguments based on sound logic, we can navigate infinity's various properties. In doing so, a whole new world of mathematics is opened, leading to even newer ideas in mathematics.

The exploration of the infinite we will try to resemble in this part led to a more robust mathematics in many ways. From providing a more rigorous basis for mathematics to inspiring the development of many other fields, the paradoxes of infinity are a testament to the power of paradoxes in guiding mathematics' progress.

But the infinite is not just one thing. As we will soon discover, there is in fact an infinite hierarchy of infinities. Also, infinity can manifest itself in many ways. There is the geometrical infinity, the algebraic infinity, the infinitely large, the infinitely small, etc. Despite all being different in key ways, one thing unites all these concepts — the human struggle in making sense of them. We will try to discover humanity's journey of overcoming this struggle ourselves through a mathematical, historical, and philosophical take at a set of paradoxes relating to the infinite.

In this part, we will attempt to boil down the most crucial of explorations in regards to the infinite. The content discussed in this part could easily span multiple volumes each thousands of pages long, but we will attempt to keep a concise account of these adventures.

6

DOES .999...=1?

Figure 28: What a better way to start a math book chapter than with a joke?

I stumbled across this question a few years ago, and was admittedly wrong at first. However, before looking at the (several) solutions this paradox had, I tried to formulate my own. I asked myself:

How is equality defined?

Well, my first instinct was to say that if the numbers on both sides are *written* the same, then equality ought to hold. That was obviously correct, but not complete. If numbers had to be *written* the same to be equal, then $.999\cdots \neq 1$. And, indeed, that is why so many struggle at first to understand this paradox.

6.1 AN ALGEBRAIC ARGUMENT

A simple algebraic argument can be done by looking at fractions and their repeating decimals. We all know that

$$\frac{1}{3} = .333\cdots$$

Multiplying the above expression by 3 gives

$$\frac{3}{3} = 1 = .999\cdots$$

A longer, perhaps more convincing method, is the following:

$$\begin{aligned}
x &= 0.999\ldots \\
10x &= 9.999\ldots && \text{multiplying by 10} \\
10x &= 9 + 0.999\ldots && \text{splitting off integer and decimal} \\
10x &= 9 + x && \text{using the definition of } x \\
9x &= 9 && \text{subtracting } x \\
x &= 1 && \text{dividing by 9}
\end{aligned}$$

Voila!

However, some mathematicians deem these arguments to be somewhat lacking. In his book *How Mathematicians Think: Using Ambiguity, Contradiction, and Paradox to Create Mathematics*, mathematician William Byers discusses how the unresolved decision on the exact meaning of the equals sign causes problems.

He describes the reaction of most undergraduate students as one of hesitation — they certainly feel that $.999\cdots$ is "very close" to 1, with some even going as far as saying that it is "infinitely close"[1], they are not quite ready to say that it is equal to 1.

While Byers describes our second argument as one with more acceptance, he still notes that it leaves the ambiguity behind the "=" symbol out there. In my experience running *@daily_math_*, I have noticed the same thing. Perhaps it was the step-up in the complexity of the argument, and the presentation of a second argument, that changed so many minds. But at the core, this explanation does not satisfy the curious student. In writing this I'm reminded of a quote by the American statistician David Blackwell, who as a Black mathematician broke so many of society's barriers put up against him and became one of the most accomplished statisticians of his time:

> *... I'm not interested in doing research and I never have been. I'm interested in understanding, which is quite a different thing.*

This alludes to our exploration of the different schools of mathematics in our fifth chapter on Gödel's incompleteness theorems. Many mathematicians to this day take the formalist approach when asked what they do, and describe their work as simply theorem-hunting and proving. But most mathematicians want more than simply proving theorems — they are out for that crucial understanding too. The proofs above, as Byers observed, leave much to be explored. And if we want to *understand*, we ought to resolve this ambiguity. And that's what we will try to achieve in the next few sections!

As we delve into more and more paradoxes, we'll uncover that there are many things that we can find at the boundary and limits of reason and mathematics. The real value lies in exploring this boundary rather than just proving it exists. A

1 More on this notion later.

bit of experimentation and exploration can result in a beautiful journey of discovery for the students and academics alike.

6.2 THE CONSTRUCTION OF THE REAL NUMBERS

Back to when I first encountered this problem; I proceeded to formulate another definition for equality. Now, this was not all that rigorous, but it was on the right track.

> Two numbers, a and b, are equal if there does not exist some c such that it lies *between* them.

Upon further investigation, I found a more rigorous version of this argument: **Dedekind cuts**. Dedekind cuts are simply a method to construct the real numbers from the rationals. They define each number into two sets, A and B, such that all elements of A are less than all elements of B, and A contains no greatest element. In other words, A contains every rational number less than the cut and B contains every rational number greater than or equal to the cut. This is because the set B may not have a smallest element in the rationals. If B does have a smallest element among the rationals, the cut corresponds to that rational. If it does not, the cut defines an irrational number that "fills the gap" between A and B, as seen in figure 29.

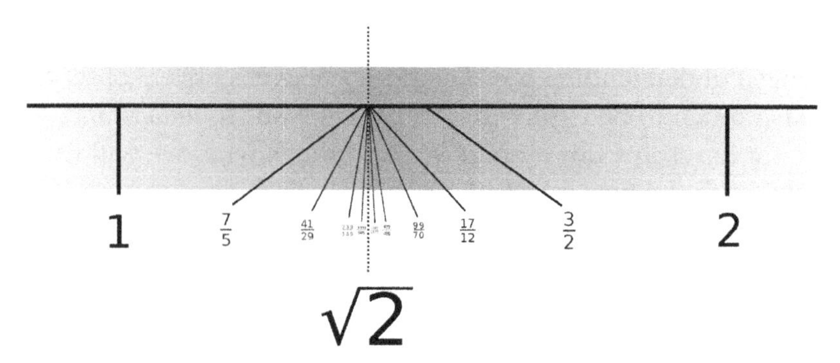

Figure 29: Dedekind cut defining $\sqrt{2}$.

Using this Dedekind definition with a bit of rigor, one can easily translate the loose notion of equality into a more solid one and prove that $.999\cdots = 1$.

6.3 GEOMETRIC SERIES

Now onto more analytical proofs. Consider writing $.999\cdots$ as:

$$.999\cdots = \frac{9}{10} + \frac{9}{100} + \frac{9}{1000} + \cdots$$

This is simply a geometric series with common ratio $\frac{1}{10}$. Therefore,

$$.999\cdots = 9\sum_{n=1}^{\infty}\left(\frac{1}{10}\right)^n$$

Using the geometric series formula,

$$\sum_{n=1}^{\infty} r^n = \frac{r}{1-r}$$

We have:

$$\begin{aligned} .999\cdots &= 9 \cdot \frac{\frac{1}{10}}{1-\frac{1}{10}} \\ &= 9 \cdot \frac{1}{9} \\ &= 1 \end{aligned}$$

Thus,

$$.999\cdots = 1$$

This proof appears as early as 1770 in Leonhard Euler's *Elements of Algebra.*

6.4 ALTERNATE NUMBER SYSTEMS

It is worth mentioning that the $.999\cdots = 1$ is a convention that does not hold up universally, similar to how in some systems dividing by "zero" actually gives meaningful results. We see that mathematics is varied in this regard, and we further understand how alternate frameworks of mathematics could emerge to replace the notions we hold so dearly now. As a matter of fact, many of these alternate number systems find their way into the arguments about the convention that $.999\cdots = 1$. Subconsciously, many of these systems "make sense" in some way that is fascinating to study. Although they may pervade mathematics teaching in a rather subtle way, they were never explicitly taught.

6.4.1 *Ultrafinitism*

Ultrafinitism rejects concepts dealing with infinity such as the infinite number of trailing 9's after the decimal point in $.999\cdots$. To the ultrafinitist, only a fixed, finite number of decimal digits in a number is meaningful. This may capture the view of some students when approaching this paradox, as many students seem to think that $.999\cdots$ ought to have a "last digit." However, this philosophical position enjoys very little support in the mathematical community and is not fully developed.

6.4.2 *Infinitesimals and Hyperreal Numbers*

What's the opposite of an infinite number? An infinitesimal! They are numbers that are closer to 0 than any real number, but are not equal to zero. In standard analysis, infinitesimals do not exist. But in nonstandard analysis, these infinitesimals exist and the whole of calculus can be reformulated within the framework of infinitesimals. Using this framework,

$$.999\cdots \neq 1.$$

Mathematician and renowned author Ian Stewart characterizes this interpretation as an "entirely reasonable" way to justify the intuition of so many students that "there's a little bit missing" from 1 in $.999\cdots$.

Now we see that the equality of $.999\cdots$ and 1 is not something arbitrary, but a direct consequence of the rules we set for math. The key principle here is the *Archimedean property*, which states that a certain structure has no infinitely large or infinitely small elements. This is an advanced notion that is used in the advanced mathematical field of abstract algebra, but it essentially means that there are no infinitesimals in the real number system. We can extend this real number system into a larger system such as the *hyperreals*, which contain infinitesimals, and obtain a whole new mode of analysis for the entirety of mathematics, not just $.999\cdots = 1$.

7

HILBERT'S HOTEL PARADOX

We might as well just start with the paradox since it's very simple to phrase.

> Hilbert's hypothetical hotel, due to Hilbert's mathematical genius, found a way to have a countably infinite number of rooms, numbered $1, 2, 3 \cdots$. Business is booming and the hotel's rooms are all occupied. Can the hotel accommodate any new guest(s)?

Well, the hotel is completely full, so the answer ought to be a very hard no, right? It turns out the answer is yes — we can even manage to fit an infinite amount of additional guests! The answer lies in the subtle (well, they are very apparent when one considers infinity more closely) differences in what operations we can treat infinite numbers with.

This paradox, similar to many others, shows us how our normal conceptions of arithmetic don't quite translate to the world of the infinite. As counter-intuitive as this paradox seems, it's at the heart of early developments in set theory. The person who was behind this paradox is David Hilbert, someone who we met earlier in the previous part of the book. Hilbert was a staunch advocate of Cantor's ideas on infinity, and introduced this paradox in a lecture given in 1924 named "Über das Unendliche," or "About the Infinite."

David Hilbert was a German mathematician known for being one of the last universalists in mathematics, i.e. excelling in and doing considerable work in several mathematical disciplines. He is known for being one of the founders of mathematical logic. His 23 unsolved problems were influential in guiding the modern trajectory of mathematics.

Figure 30: David Hilbert (1862 - 1943)

Hilbert's paradox was later popularized by the book *One Two Three... Infinity* by American physicist George Gamow. Hilbert's paradox was an important stepping stone that widely popularized the realization that the rules applying to finite numbers, more specifically those of arithmetic, do not extend to the infinite. The mathematical ideas behind this paradox weren't anything new — although many creative approaches were taken, the essential ideas were cemented by Cantor's theories. Nonetheless, this paradox, perhaps owing to using a concrete object like a hotel instead of abstractions like sets of numbers, widely popularized the notions behind Cantor's theory.

To make an overly simplified conclusion behind this paradox, it is that

$$\infty + 1 = \infty + \infty = \infty.$$

Let's talk about exactly how our overly simplified equation is correct[1].

1 As mentioned before, one cannot treat infinity as a number, hence the equality symbol here is partially incorrect. But as we explained before, ambiguity

7.1 WHAT DO OUR METHODS TELL US?

A naive approach would be to simply apply the pigeonhole principle, which states:

> if one has m objects to put into n boxes (in this case, m guests into n rooms) such that $m > n$, then at least one box has more than 1 object in it.

If we were to use this, we would get a contradiction leading us to believe that we cannot accommodate any new guests!

But, in fact, we can accommodate an additional guest, or any finite number of additional guests! The most counterintuitive fact is that we can even fit in an additional infinite amount of guests![2]

7.2 RETHINKING THE PROBLEM

7.2.1 *Finite Guests*

Okay, let's try the easiest scenario first. How can we accommodate one additional guest if all of the rooms are filled? It's quite simple actually!

Move each person already in the hotel from their room to the next. In mathematical terms, if a guest already accommodates room number n, then we can simply move them to room $n+1$. Since the hotel has infinite rooms, there will always be a room with number $n+1$. After this process, room 1 is vacant, and we have a space for our guest. We can easily extend this to any *finite* number of new guests, k, by replacing our rule of $n \to n+1$ to $n \to n+k$.

---

sometimes is the catalyst of great exploration, so we'll try to determine in what sense that equation is "true."

2 Countably infinite, that is. We will take a look at what that exactly means in the next chapter.

7.2.2 *Infinitely Many Guests*

But how are we going to accommodate an infinite amount of new guests? Again, this is a simple fix! Following the logic from the simple finite case, we can move each guest in room n to room $2n$, thereby clearing up all the odd rooms for our new guests. Since we have an infinite number of hotel rooms, $2n$ will always be a valid room number!

This is the crucial concept: we have a *pairing function*. In this case, it is mapping $n \to 2n$. In our finite case, it was the mapping $n \to n + 1$. The beauty behind this method is, since n has no bound, the logic is sound. Paradoxical, yes, but sound!

We aren't done here, though, let's take it up a notch.

7.2.3 *Infinitely Many Buses With Infinite Guests*

Let's say we want to have an infinite number of buses, each carrying an infinite number of new guests. Can we accommodate all of these guests as well? Think of what we did before and be creative! There are many solutions, or *pairing functions*, for this problem. We will take a look at a few such solutions.

Powers of Prime Numbers

Let the ith prime number be denoted by P_i. Then, one possible solution is assigning the kth member of the nth bus to the $(P_n^k)^{\text{th}}$ room. Notice that since we are only using prime numbers, each bus-rider will get a unique room. Paradoxically, this will leave all numbers with multiple prime factors as empty, thereby leaving an infinite amount of rooms empty. Yes, this is very counterintuitive.

Prime Factorization

The key to this method is recognizing that each number's prime factorization is unique. For example, $96 = 3^1 \cdot 2^5$ and cannot be factorized in another way with prime numbers. This is the *fundamental theorem of arithmetic.*

After recognizing that, we can assign the kth member of the nth bus to the $(2^k 3^n)^{\text{th}}$ room. But notice that we can extend this to higher levels by multiplying more prime numbers. For example, suppose that there is another group of infinitely many buses, each with infinitely many guests, every night. Then, on the ith night, we can assign the kth member of the nth bus to the $(2^k 3^n 5^i)^{\text{th}}$ room. We can keep doing this for higher and higher levels of nesting, each time adding a new power of a prime number.

Notice that, similar to the previous method, we have also left an infinite amount of rooms empty! It is worth noting that we can extend this thought-experiment into even higher levels of nesting as well.

7.3 THE NEW MATHEMATICS

We just found, through a simple thought-experiment, that the properties of the infinite are quite different from those of the finite. This paradox can be understood by using Georg Cantor's theory of transfinite numbers, and is a very useful exercise in introducing Cantor's fundamental ideas. What are those ideas? We will explore them in the next chapter.

8

ARE SOME INFINITIES BIGGER THAN OTHERS?

Are they? It seems from our last chapter that all infinities are virtually the same size. It seems like adding to the infinite never changes its size. Maybe it does not make any sense to talk about greater and less than in the context of infinite sets. This is certainly what the Italian polymath Galileo Galilei thought.

Before we delve into the work of Georg Cantor on infinite sets — which shapes much of our understanding today about these objects— it is worth noting an early observation by Galileo. *Galileo's paradox* states that, even though most natural numbers are not squares, the size of the set of square numbers is equal to that of the entirety of the natural numbers. This is one of the earliest uses of the concept of *one-to-one correspondence*. We will see how this comes to life in the work of Georg Cantor.

We have referenced Cantor and his work many times in this book without pausing to delve deeper, but it's time to change that. The exploration we did in the previous chapter indicates that all infinities are essentially equivalent, and that was the very belief that permeated the mathematical community until the work of Cantor. In his first paper on set theory, "On a Property of the Collection of All Real Algebraic Numbers," Cantor proved that there is in fact more than one kind of infinity.

Georg Cantor was a German mathematician who developed the modern foundational theory of mathematics — set theory. He is recognized as one of the greatest mathematical minds of the 19th and 20th centuries, and his work can be seen incorporated into virtually every field of mathematics. His work has also generated much philosophical interest.

Figure 31: Georg Cantor (1845 - 1918)

At first, he referred to *countable* infinity and *uncountable* infinity, the latter of which was not believed to exist until Cantor's rigorous proof. An uncountable infinity, as the name suggests, is a type of infinity that *can not* be counted — like the real numbers. Cantor proved the incredible result of the uncountability of the reals through the method of *diagonalization*, which you might recall from our chapter on Gödel's incompleteness theorems.

Between 1879 and 1884, Cantor published a series of articles that together formed the basis of his revolutionary ideas. As is expected with any ideas regarding infinity, there was considerable opposition. The German mathematician Leopold Kronecker, from the intuitionist camp of mathematical philosophy, led this opposition. He was a *finitist* who only believed that mathematical concepts are valid if they could be constructed in a finite number of steps from the natural numbers. His criticism was fierce, publicly condemning Cantor and calling him a "corrupter of youth." Besides Kronecker, many other mathematical figureheads such as Poincaré, Brouwer, and Weyl objected to Cantor's ideas.

The criticism wasn't only mathematical. Philosophers like Ludwig Wittgenstein raised serious philosophical objections. Even some theologians challenged Cantor's work as they saw it as a challenge to the "absolute infinite" associated with God. They went as far as to accuse Cantor, a devout Lutheran, of being a pantheist!

However, this heavy criticism was later met with heavy praise. As mathematics progressed, mathematicians and the world at large began to see the value in Cantor's ideas. Now, the entire foundation of mathematics — ZFC set theory — is built on Cantor's early ideas about set theory. David Hilbert was one of the most active supporters of Cantor's ideas, and defended them by declaring:

> *No one shall expel us from the paradise that Cantor has created.*

8.1 SOME SIMPLE OBSERVATIONS

Take the set of all natural numbers and the set of all even numbers.

$$\mathbb{N} = \{1, 2, 3, \cdots\}, \ \mathbb{E} = \{2, 4, 6, \cdots\}$$

Both of these sets obviously have an infinite size, or *cardinality*. Relatively, though, is $\mathbb{E}$ smaller than $\mathbb{N}$?

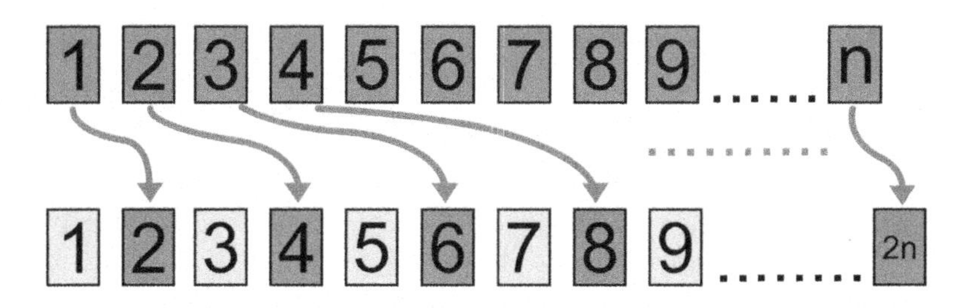

Figure 32: Pairing the natural numbers and the even numbers. Source: Wikimedia

Notice that $\mathbb{E}$ only has half the numbers in $\mathbb{N}$, so $\mathbb{E}$ ought to be smaller, right? Surprisingly, however, these sets are of equal size. The way we go about proving that is through a *bijection* of sets, which we can show through figure 32.

Essentially, if two sets can be put into a bijection or a one-to-one correspondence, then they are of equal size! This is the same observation that Galileo made. Now, what about the set of all rational numbers?

$$\mathbb{Q} = \left\{\frac{1}{1}, \frac{1}{2}, \frac{1}{3}, \ldots, \frac{2}{1}, \frac{2}{2}, \frac{2}{3}, \ldots, \frac{3}{1}, \ldots\right\}$$

Surprisingly, the set of all rational numbers is *also* the same size as that of the naturals! The way to establish a bijection here is much more tricky, however.

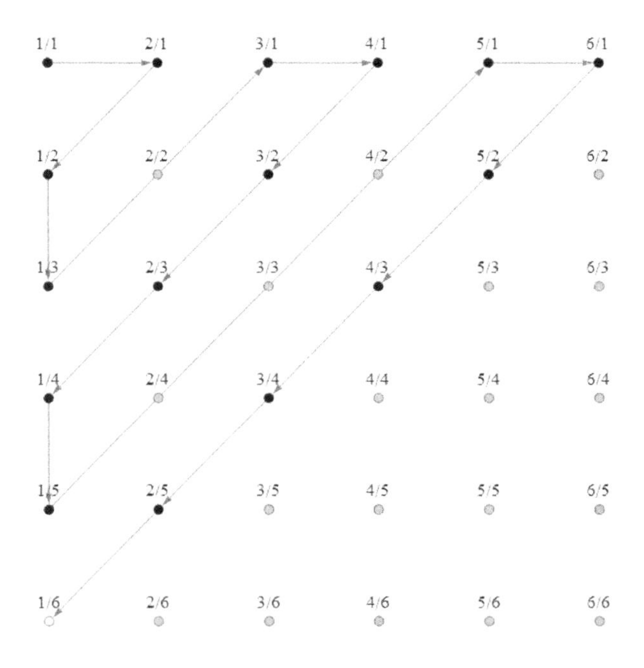

Figure 33: The arrows indicate the order of assigning the naturals to the rationals. This can be easily extended to include negative rationals (left as a challenge for the reader).

We just saw that an infinite set may have the same cardinality as a proper subset of itself, three times! This is mind-blowing news. But so far, all these infinities seem to be of the same size, so are *all* infinities of the same size? Does it make no sense to compare sizes of different infinite sets?

8.2 DIAGONALIZATION

Let's take a look at the set of all real numbers, $\mathbb{R}$. This is a *bigger* infinite set in the sense that it contains all the elements of the set of all rational numbers as well as more elements.

Can we set up a certain bijection such that we prove that its cardinality is equal to that of the set of natural numbers? Cantor thought not. As a matter of fact, we'll prove that even the set of real numbers between 0 and 1 is bigger than the set of all naturals! Imagine assigning each natural number to a real number, as the following:

$$
\begin{aligned}
1 &\to .\boxed{3}14829\cdots \\
2 &\to .2\boxed{4}5094\cdots \\
3 &\to .98\boxed{6}574\cdots \\
4 &\to .765\boxed{7}48\cdots \\
5 &\to .0295\boxed{8}0\cdots \\
&\vdots
\end{aligned}
$$

Take the first digit in the number corresponding to one, the second digit in the number corresponding to two, etc, and construct a *new* number by adding 1 to each digit that we have. This new number must not be in our list because it is different from each number originally in our list by at least one decimal. So, we have just constructed a whole new number not in our list that matched natural numbers to real numbers between 0 and 1. Therefore, there are more numbers between 0 and 1 than natural numbers!

8.3 IN POPULAR CULTURE

In his novel *The Fault in Our Stars*, John Green writes:

> There are infinite numbers between 0 and 1. There's .1 and .12 and .112 and an infinite collection of others. Of course, there is a bigger infinite set of numbers between 0 and 2, or between 0 and a million. Some infinities are bigger than other infinities... I cannot tell you how grateful I am for our little infinity. You gave me forever within the numbered days, and I'm grateful.

Apparently, yes, some infinities are bigger than others. As a matter of fact, there are an *infinite* number of sizes of infinity. We can talk about these sizes through the language of *transfinite numbers*.

8.4 TRANSFINITE NUMBERS

Okay, by now, we know that infinity is not a number but rather a concept. Our system of arithmetic does not work on infinity only because arithmetic was designed to work with numbers rather than concepts.

But what if I told you that ∞ can be a concept codified in "numbers" as well?

Transfinite numbers, like so much of our understanding of infinity, arose from the early study of set theory done by Georg Cantor.

The smallest infinity is the size of the set of natural numbers. It is defined as $\aleph_0$, pronounced "aleph-null." The next, larger infinity can be generated by taking the *power set* — or the set of all possible subsets— of the natural numbers. This is the set of all real numbers: $\aleph_1$. Or is it?

8.4.1 *The Continuum Hypothesis*

The continuum hypothesis is a fascinating assumption we make about sets and their cardinality. It essentially states that there is no cardinal number between the cardinality of the reals ($\mathfrak{c}$) and the cardinality of the natural numbers, i.e. $\aleph_1 = \mathfrak{c}$. This hypothesis was in fact one of the first statements shown to be independent, i.e. could not be proven or disproven, from the axioms of ZFC which mathematics operates on.

8.4.2 *Even Larger Infinities?*

We can continue this process of generating larger infinities without end. All we need to do is take the power set of our $\aleph_n$ to generate an $\aleph_{n+1}$. And just like there is no largest number, there is no largest infinity!

9

GRANDI'S SERIES

Series are simply infinite sums, for example:

$$\sum_{n=1}^{\infty} \frac{1}{n^2} = \frac{1}{1^2} + \frac{1}{2^2} + \frac{1}{3^2} + \cdots$$

A curious fact about this series is that it equals $\frac{\pi^2}{6}$, which is discussed at length and derived in my book *Advanced Calculus Explored.* Now, this series *equals* something in the traditional sense that we define equality. To be more rigorous, we simply say that this series *converges* to $\frac{\pi^2}{6}$. To put it loosely, "converging" is simply the word for "getting closer and closer to" a specific number. But, not all series converge!

Perhaps the most well-known series that does not converge, or diverges, is the *harmonic series*:

$$\sum_{n=1}^{\infty} \frac{1}{n}.$$

This series approaches ∞. And that is what most people tend to think of when a series is said to *diverge*. Nonetheless, that isn't necessarily the case. Diverging is defined through a logical negative: *a series diverges if it does not converge.* So, if a series does not approach any specific number, it diverges.

An example of this would be **Grandi's series:**

$$1 - 1 + 1 - 1 + 1 - 1 + \cdots = \sum_{n=0}^{\infty} (-1)^n$$

What does this series equal? Let's dive into various people's attempts to figure it out.

9.1 HISTORY

9.1.1 *Grandi*

Grandi's series was first studied by Guido Grandi, who gave an extensive treatment of it in 1703.

Guido Grandi was an Italian mathematician, philosopher, and priest. Initially a professor of theology and philosophy, he took interest in mathematics later in his life. He is primarily known for his studies on curves such as the rose curve and the witch of Agnesi, the latter of which helped introduce calculus to Italy.

Figure 34: Guido Grandi (1671 - 1742)

Grandi took an interesting approach, characteristic of his time not only mathematically but philosophically. Grandi observed that one can order the operations present in the series in different ways to give different results:

$$(1-1)+(1-1)+(1-1)+\cdots=0$$

$$1+(-1+1)+(-1+1)+\cdots=1$$

Grandi took this mathematical result and ran away with it, claiming with a strong religious overtone that

> *By putting parentheses into the expression 1 - 1 + 1 - 1 + ⋯ in different ways, I can, if I want, obtain 0 or 1. But then the idea of the creation ex nihilo is perfectly plausible.*

Nonetheless, Grandi did not believe the series' value to be either 1 or 0. He claimed that the series actually equals $\frac{1}{2}$. Grandi essentially used the series expansion

$$\frac{1}{1+x} = 1 - x + x^2 - x^3 + \cdots, \tag{8}$$

plugged in $x = 1$ and voila! The value of $\frac{1}{2}$ was arrived at. Grandi was a naughty mathematician in this regard, ignoring intervals of convergence and plugging away at general formulas. Only if mathematics was that simple! In fact, the series expansion in (8) only holds for $|x| < 1$. Grandi also claimed that since the series can be shown to equal 0 *and* $\frac{1}{2}$, the world could be created out of nothing. A Grand(i) conclusion indeed.

Grandi offered another approach to arrive at his answer of $\frac{1}{2}$ several years later:

> *Two brothers inherit an expensive gem from their father, whose will forbids them to sell it. They then agree that it will reside in each other's residences on alternating years. If this agreement is held indefinitely across generations, then the two families will each have half possession of the gem, even though the gem alternates between residences infinitely many times.*

9.1.2 *Leibniz*

Gottfried Leibniz, who is known for co-inventing calculus with Isaac Newton, criticized Grandi's second method but thought the argument stemming from the power series expansion of $1/(1+x)$ was valid. Following from his reasoning in calculus and limits, Leibniz claimed that since (8) holds for $|x| < 1$, it is valid to take its value at $x = 1$.

9.1.3 *Euler*

Euler took issue with some of Leibniz's ideas, perhaps as a consequence of the relatively more rigorous underpinnings he had compared to Leibniz. After looking at other diverging series such as $1+2+4+\cdots$, Euler argued that:

> *Yet however substantial this particular dispute seems to be, neither side can be convicted of any error by the other side, whenever the use of such series occurs in analysis, and this ought to be a strong argument that neither side is in error, but that all disagreement is solely verbal. For if in a calculation I arrive at this series* 1 - 1 + 1 - 1 + 1 - 1 *etc. and if in its place I substitute* 1/2*, no one will rightly impute to me an error, which however everyone would do had I put some other number in the place of this series. Whence no doubt can remain that in fact the series* 1 - 1 + 1 - 1 + 1 - 1 + *etc. and the fraction* 1/2 *are equivalent quantities and that it is always permitted to substitute one for the other without error. Thus the whole question is seen to reduce to this, whether we call the fraction* 1/2 *the correct sum of* 1 - 1 + 1 - 1 + *etc.; and it is strongly to be feared that those who insist on denying this and who at the same time do not dare to deny the equivalence have stumbled into a battle over words. But I think all this wrangling can be easily ended if we should carefully attend to what follows...*

9.1.4 *Other Approaches*

In a message sent to the Italian mathematician Joseph Langrange, Jean-Charles Callet notes that:

$$\frac{1+x}{1+x+x^2} = 1 - x^2 + x^4 - x^6 + \cdots$$

Which, if $x = 1$ is plugged in, gives the series value of $\frac{2}{3}$. We can now see the danger of using series outside of their valid limits! Arguably, we could construct an infinite number of series that could correspond to Grandi's series, and each could give different values. Perhaps a stronger argument is needed.

9.2 RESOLUTION

As we will see in the next chapter, various methods were developed in the 19th century that assigned specific finite values to divergent series. They can be defined rigorously, and often have much use inside and outside of mathematics. This helps us avoid the *Eilenberg–Mazur swindle* which is the pitfall of using heuristic methods to play with infinite divergent series, which Grandi and so many others did. Although this concept is an advanced one, discussing pitfalls of some methods in geometric topology and abstract algebra, it can be more or less replicated by the following argument on Grandi's series:

$$\begin{aligned} 1 &= 1 + (-1 + 1) + (-1 + 1) + \cdots \\ &= 1 - 1 + 1 - 1 + \cdots \\ &= (1 - 1) + (1 - 1) + \cdots = 0 \end{aligned}$$

which is a contradiction. Virtually all of the widely-used methods to assign values to divergent series — including Abel, Borel, and Cesaro summation — give Grandi's series a value of $\frac{1}{2}$.

9.3 LEGACY

Grandi's series and its rich history symbolized the confusion on divergence in the mathematical community that ruled discussions for centuries. It wasn't until the mathematician Cauchy that modern mathematics with all of its rigor was developed. As the mathematician Judith Grabiner once stated:

> *"...Since Newton the limit had been thought of as a bound which could be approached closer and closer, though not surpassed. By 1800, with the work of L'Huilier and Lacroix on alternating series, the restriction that the limit be one-sided had been abandoned. Cauchy systematically translated this refined limit concept into the algebra of inequalities, and used it in proofs once it had been so translated; thus he gave reality to the oft-repeated eighteenth-century statement that the calculus could be based on limits."*

However, after calculus and analysis became more rigorous, the emphasis on convergence and divergence became almost too great to bear. In modern mathematics education, series are often presented without the rich history behind them or their applications. The student then always wonders "Why are we doing this?" The preoccupation with determining convergence, but not the actual sum, makes the study of series so much more artificial, and often times, pointless to many students. The further study of the analytical tools we delved into above would be an excellent addition to the mathematical curriculum, as they are crucial in the mathematical sciences.

9.3.1 *Thomson's Lamp*

Thomson's lamp is a thought experiment seemingly related to Grandi's series. It was devised in 1954 by the British philosopher James F. Thomson, who used it to analyze the possibility of a supertask (See the chapter on Zeno's paradoxes).

However, in his original 1954 paper, Thomson wanted to differentiate supertasks from their series analogies. He writes:

> *Then the question whether the lamp is on or off ... is the question: What is the sum of the infinite divergent sequence*

$$+1, -1, +1, \cdots ?$$

Now mathematicians do say that this sequence has a sum; they say that its sum is 1/2. And this answer does not help us, since we attach no sense here to saying that the lamp is half-on. I take this to mean that there is no established method for deciding what is done when a super-task is done. ... We cannot be expected to pick up this idea, just because we have the idea of a task or tasks having been performed and because we are acquainted with transfinite numbers.

10

1 + 2 + 3 + 4 + ...

The claim

$$1+2+3+4+\cdots=-\frac{1}{12}$$

has become one of the most notorious "memes" within the mathematical community. It is also one of the most mind-blowing claims that are (in a sense) true.

This is similar to Grandi's series — simply another divergent series with a peculiar result. But, why has it generated so much more attention that Grandi's series? Notice that, unlike Grandi's series, $1+2+3+4+\cdots$ trivially tends to ∞ while Grandi's series did not approach any number and is simply alternating between 0 and 1. Assigning $\frac{1}{2}$ as Grandi's series value was a bit of an uncomfortable claim, but definitely not as outrageous as assigning the above series, which approaches a positive infinity, to a fractional negative value. Before we delve in, however, we should make one thing clear and remind ourselves that mathematics is ever-evolving, and series like this one caused it to evolve new methods of assigning values to these divergent series. Why? Infinite answers often were not satisfactory.

In many fields where this sum is encountered, such as in complex analysis, quantum field theory, and string theory, an infinite answer does not make sense. Instead, if we were to say

$$1+2+3+4+\cdots=-\frac{1}{12}$$

and assign the series a value of $-\frac{1}{12}$, then we would have correct results! The subtle notion behind this is that we are not claiming that $1+2+3+4+\cdots$ is actually equal to $-1/12$, but only *assigning that value* to the series.

But how? How could the sum of positive whole numbers equal a negative fraction? This is perhaps one of the most striking of paradoxes presented in this book. It turns out, through some clever methods — even more so than the ones introduced in the previous chapter — we can assign that value to the sum. Let's jump right in!

10.1 SOME APPROACHES

The series was mentioned by Euler in 1760 alongside another divergent series: $1+2+4+8+\cdots$. Euler hints that these series have finite negative sums, but does not go into much detail about the series $1+2+3+4+\cdots$ and its value. It was not until Ramanujan did this series become popularized.

10.1.1 *Ramanujan's Approaches*

Srinivasa Ramanujan presented two methods to reach the value of $-\frac{1}{12}$.

Srinivasa Ramanujan was an Indian mathematician. Even though he had no formal training in mathematics, Ramanujan was a brilliant mathematician. Initially failing to garner any attention for his work, the English mathematician G.H. Hardy arranged for him to travel to Cambridge in 1918. During his short life, Ramanujan managed to compile almost 4000 mathematical results, mainly identities.

Figure 35: Srinivasa Ramanujan (1887 - 1920)

Let's employ Ramanujan's heuristic approach. Begin with writing the sum:

$$\begin{aligned} S &= 1+2+3+4+5+6+\cdots \\ 4S &= \quad\ 4 \quad\ +8 \quad +12+\cdots \\ S-4S &= 1-2+3-4+5-6+\cdots \end{aligned}$$

Now, we recall the power series of the function $f(x) = \frac{1}{(1+x)^2}$:

$$\frac{1}{(1+x)^2} = \sum_{n=1}^{\infty}(-1)^{n-1}nx^{n-1} = 1-2x+3x^2-4x^3+\cdots$$

If we were to forget rigor and just plug in $x = 1$ without checking for convergence, we end up with

Another way of finding the constant is as follows —
Let us take the series 1+2+3+4+5+&c. Let C be its constant. Then
$$C = 1+2+3+4+\&c$$
$$\therefore 4C = 4 + 8 + \&c$$
$$\therefore -3C = 1-2+3-4+\&c = \frac{1}{(1+1)^2} = \frac{1}{4}$$
$$\therefore C = -\frac{1}{12}.$$

Figure 36: Excerpt from Ramanujan's first notebook.

$$-3S = \frac{1}{4}$$
$$S = -\frac{1}{12}$$

Voila!

A more sophisticated and rigorous way is *Ramanujan summation.* As we saw in the previous chapter, there is sometimes more than one way to sum series. Besides Ramanujan summation there exist many different definitions of summation that yield and do not yield results for specific functions. Although these types of summations do not resemble the traditional sense of the word, they are useful in areas such as quantum field theory.

10.1.2 *Zeta Function Regularization*

Another method we can use is through the zeta function. The zeta function is a *special function,* i.e. a function that appears so often in mathematics and science it was useful to give it a name and study its properties. It is used in fields from analytic number theory in the discussion of primes to statistical mechanics in the discussion of Bose-Einstein condensates (an ultra-cold state of matter). What exactly is this function, though?

The zeta function can be defined as the *analytic continuation* of the series:

$$\zeta(s) = \sum_{n=1}^{\infty} \frac{1}{n^s}$$

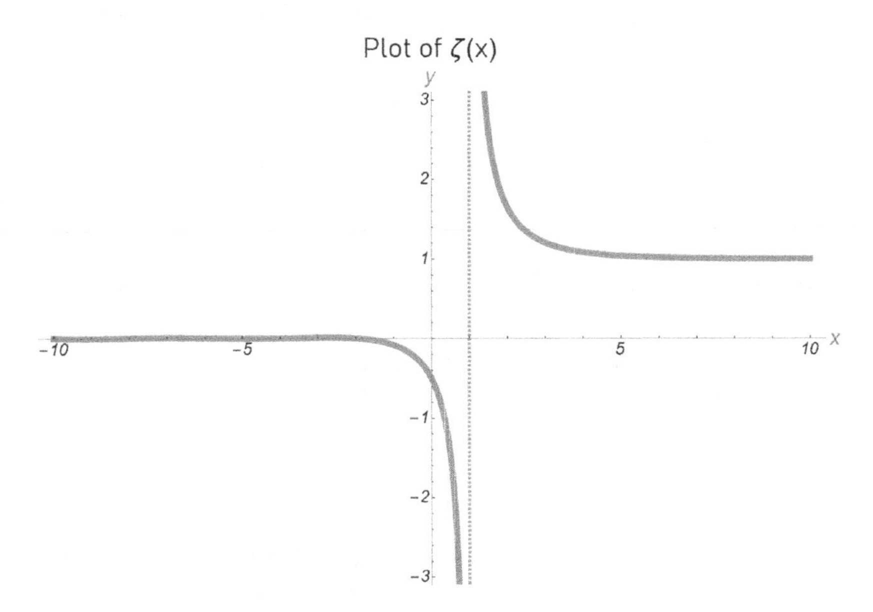

Figure 37: Notice that the zeta function diverges at $x = 1$

For $s = 1$, the zeta function is simply the harmonic series. For $\text{Re}(s) > 1$, the zeta function is actually convergent[1]. For even values $2, 4, \cdots$ the function produces an interesting set of results involving powers of π as seen in (9). I have done a lengthy analysis of this function and its applications in my previous book *Advanced Calculus Explored*. Graphing this function on the Cartesian plane gives the plot in figure 37.

The beauty in this method of evaluating the series is the method of *analytic continuation*, a very important technique in higher mathematics. The essence of this technique is looking at a function which only "makes sense," in this case converges, on a particular interval and then analytically extending that interval. One can then define the zeta-regularized sum of $1 + 2 + 3 + 4 + \cdots$

1 "Re" denotes the real part of a number. The zeta function is often studied on the complex plane, and a million-dollar problem, the Riemann hypothesis, concerns its behavior on the complex plane.

to be $\zeta(-1)$. We can then evaluate $\zeta(-1)$ through a variety of methods to yield the value of $-1/12$.

10.2 APPLICATIONS IN PHYSICS

Quantum field theory (QFT) attempts to explain physics through the combination of classical field theory, special relativity, and quantum mechanics. It treats particles as simply excitations of their corresponding fields. The series $1+2+3+4+\cdots$ finds application in the *Casimir effect* in QFT.

The Casimir effect is simply the description of the force that arises from a quantized field. The classic example is to think of two uncharged, conductive plates placed very close to each other, a few nanometers apart let's say. In classical physics, these plates would not exert any electromagnetic force on each other because of the lack of an external field. When this field is studied under QFT, however, the plates affect the virtual photons which constitute the field, and as a result generate a net force. In the derivation of this effect, one runs into the sum $1+2+3+4+\cdots$ and the result of $-1/12$ produces consistent results.

This series also finds application in string theory, the most promising candidate for a theory of everything of physics. It's puzzling how seemingly arbitrary definitions of the sum of divergent series enjoy such widespread applicability in physics!

Part IV

GEOMETRY THAT DEFIES INTUITION

Geometry is the archetype of the beauty of the world.

— Johannes Kepler

Oh, geometry! The oldest mathematical discipline, the force that drove the exploration of mathematics for millennia, and the universal framework in which we study nature.

Why is geometry so seemingly universal? Well, it can be easily explained through our biology. We are very much visual, hands-on creatures. Fields like algebra and probability, although governing much of our lives, were largely undiscovered until much later in humanity's development because of their abstract nature. Geometry, on the other hand, was *very* early to the race, coming about 2000 years before algebra and probability solidified into proper mathematics fields. The study of forms and shapes constitutes much of our development as children. Geometry is then just a natural extension of the learning process we all go through!

One of the earliest *geometers* was Pythagoras — famous for his theorem on right triangles. Euclid then followed with his axiomatized system of geometry and one of the most influential texts in history: *The Elements* — widely referred to as the *most successful and influential textbook ever written*, and for good reason. The book is only second to the Bible in terms of the number of editions published since the text's first printing in 1482! That's one astronomical accomplishment.

Euclid of Alexandria was a Greek mathematician known for being the father or founder of the discipline of geometry. His historic book on geometry, *The Elements*, was the main textbook used in math curricula worldwide up until the 19th century. He also worked on number theory and physics.

Figure 38: Euclid (c. 325 BC - 270 BC)

Geometry also arose independently in India, with various texts describing geometric constructions as early as the 3rd century BC. The Indian and Greek traditions were continued and built upon by the mathematicians of the Islamic golden age. By the 17th century, through the early work of Islamic mathematicians and the new work of European mathematicians such as René Descartes and Pierre de Fermat, geometry took a much more analytic turn and began to expand beyond its Greek origins rapidly.

In the 17th century, two important developments occurred within the realm of geometry. The first was due to Descartes, and was the creation of analytic geometry or "coordinate geometry," which gave geometry an algebraic basis. Fermat was also key in this development. Second was the development of projective geometry[2] by Girard Desargues. This revolutionized geometry and led to entirely new insights into the world of forms and shapes.

But, nothing quite compared to the geometric revolution of the 19th century. Like we discovered in the chapter on Gödel's incompleteness theorems, non-Euclidean geometries were developed in this period. The two master geometers of this period, Bernhard Riemann and Henri Poincaré, each developed their own branch of this new geometry. Riemann used his analytic tools and introduced such concepts as Riemann surfaces and others that were instrumental to the development of Einstein's general relativity. Poincaré, on the other hand, founded algebraic topology. This led to a revolution in how we see shapes, forms, and space, and made geometry a much richer field that found applications in diverse fields, both within and outside mathematics. Through the lens of paradox, let's discover a bit about the more "exotic" geometry that came after Euclid.

---

2 Projective geometry is the study of properties of shapes under projections.

11

GABRIEL'S HORN

Cubes, spheres, tetrahedrons, and all other common shapes have finite surface areas *and* volumes. We can also have "infinite" objects — for example, we can think of an "infinite" cube such that it has an infinite surface area as well as an infinite volume.

But what about a shape with an infinite surface area, but a finite volume? Does that shape exist? As it turns out, Gabriel's horn is one such example.

11.1 HISTORY

Another name given for this curious object is *Torricelli's trumpet*, after the first person who studied it: Evangelista Torricelli.

Gabriel's horn alludes to the archangel Gabriel, a notable figure in the Abrahamic religions of Judaism, Christianity, and Islam, who is believed to blow the horn to announce Judgement Day. However, upon further inspection, it was not until the 14th century, centuries after the birth of all the Abrahamic religions, that the first mention of Gabriel as the trumpeter emerged. Nonetheless, the name is rather symbolic, as Judgement Day is portrayed as the bridge between the "finite," human life, and the "infinite," the afterlife.

This naming is very symbolic of the times that Torricelli lived in, as we saw in the case of Grandi's series and his attempt to incorporate the divine into his conclusions. It was not until much later that mathematics and science became the secular fields we know them as today.

Evangelista Torricelli was an Italian physicist and mathematician. Beyond Gabriel's horn, Torricelli is best known for his invention of the barometer and his advances in optics. His investigation of Gabriel's horn prompted a fierce controversy about the nature of infinity, even involving the influential English philosopher Thomas Hobbes.

Figure 39: Evangelista Torricelli (1608 - 1647)

Let Gabriel's horn be the shape made by taking the graph of $y = \frac{1}{x}$ and rotating it three-dimensionally about the $x-$axis from $x = 1$ to $x = \infty$. This shape has an infinite surface area but a finite volume.

11.2 DERIVATION

To calculate the volume of a solid of revolution, we can use the *disk method*. This method is simply the process of integrating tiny disks of width $\mathrm{d}x$ along a certain domain. If you are not familiar with calculus or forgot it, think back to our discussion in the third chapter but in 3-dimensions. The area of these tiny disks is $\pi(f(x))^2$, from the formula for the area of a circle. Therefore, their volume is $\pi(f(x))^2\mathrm{d}x$. Integrating these tiny disks over the domain then gives us that

$$V = \int_{x=a}^{x=b} \pi(f(x))^2 \mathrm{d}x$$

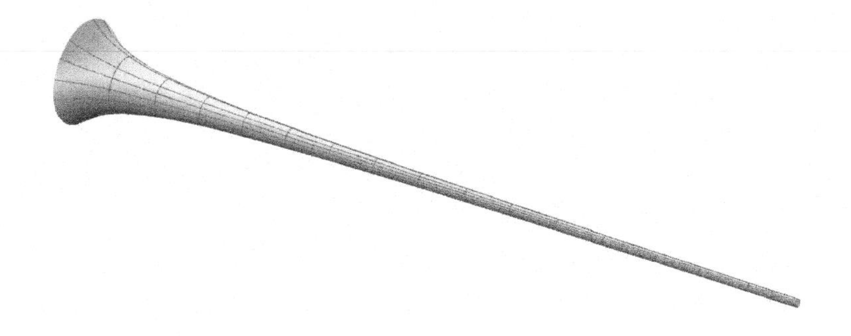

Figure 40: 3D illustration of Gabriel's horn, created with Mathematica software.

We can now do the mathematics to find out that the volume of the solid generated by revolving $y = \frac{1}{x}$ around the x−axis from $x = 1$ to $x = b$ is

$$\begin{aligned} V &= \int_1^b \frac{\pi}{x^2}\, \mathrm{d}x \\ &= \left[-\frac{\pi}{x}\right]_1^b \\ &= \pi\left(1 - \frac{1}{b}\right) \end{aligned}$$

As we take the *limit* as b tends to infinity, we see that

$$\frac{1}{b} \to 0 \implies 1 - \frac{1}{b} \to 1,$$

and the value of the volume approaches π! Therefore, we know that this shape has a finite volume, namely π cubic units.

What about the object's surface area? Well, let's think of how we calculated volume:

> Divide the shape into tiny disks with width dx and find their area. Multiplying the area and the tiny width dx together gives the volume of the disk. We then integrate along the domain, which is basically summing up all of the volumes of the tiny disks.

We can think of a similar method to find surface area. Instead of measuring the area of the disks, this time we will measure their circumference! That way, we measure the contribution of each disk to the total surface area. The circumference of a disk of radius r is $2\pi r$. Therefore, for each disk of radius $f(x)$, the circumference is $2\pi f(x)$. Again, we are dealing with disks of width dx so the area enclosing each tiny disk is *at least* $2\pi f(x)\,dx$. This is an under-estimate as if we think of one little disk, it will have an x−length of dx, but since it is at a slope, we know that its width is actually longer due to the Pythagorean theorem[1]. Summing these areas up using integration from $x = 1$ to $x = b$ gives:

$$\begin{aligned} SA &> \int_1^b 2\pi f(x)\,dx \\ &= \int_1^b \frac{2\pi}{x}\,dx \\ &= \Big[2\pi \ln x\Big]_1^b \\ &= 2\pi \ln b \end{aligned}$$

1 The actual formula for the surface area of a solid of revolution is given by $SA = 2\pi \int_a^b f(x)\sqrt{1 + (f'(x))^2}\,dx$.

However, in this case, as b tends to infinity, the surface area tends to infinity as well! We then have a shape with a finite volume but an infinite surface area. Counter-intuitive at best.

11.3 INSIGHT

When I first encountered Gabriel's horn, it blew my mind! Yet I remembered a very simple occurrence that seemed to explain it intuitively.

For me, it was Play-Doh.

Figure 41: Some Play-Doh.

I loved making snakes, and of course, the way to make snakes is by rolling the Play-Doh. If you roll the snake such that it becomes half as thick as it was before, its radius decreases by half, so its cross-sectional area decreases by one-fourth. But the volume stays the same! Since the volume is the same, the snake is now four times as long. And the surface area, which is equal to the product of the length and the diameter, has doubled. One can see that repeatedly doing this will leave the volume finite but the surface area approaching infinity!

This is similar to the case with Gabriel's horn. But, instead of being rolled, I just imagined it as a Play-Doh horn being stretched.

While making my Play-Doh snake, the volume did not change. But, the surface area always grew. Thinking of Gabriel's horn through this "stretching" can give a non-rigorous, intuitive way to understand how this is possible.

Gabriel's horn sets up the tone for the next chapter. The paradoxes in this part are unlike the others as they are much more visual. In a sense, this makes them significantly more puzzling and a result, exciting. The next paradox we will talk about is one that has generated much discussion within the mathematical community and beyond — but it will be a far cry to call it anywhere near intuitive.

12

THE BANACH-TARSKI PARADOX

Can we take a 3D ball, split it into certain sections, then reassemble it to make *two* balls of the same size as the original? Something like this:

Figure 42: A sphere is reassembled into two identically sized spheres.

Obviously not, that would be crazy! Not so fast — it turns out that we can. In the early 20th century, this is exactly what Stefan Banach and Alfred Tarski came up with. They asked the question:

> Can a ball be decomposed into a finite number of point sets and reassembled into two balls identical to the original?

They determined the answer is yes. This theorem would go on to become one of the most counter-intuitive and profound theorems in the 20th century. Let's take a deeper dive.

12.1 A USEFUL ANALOGY

However, before delving deep into the mathematics and intricacies of this paradox, let's think of a much cleaner analogy. In essence, one can think of the Banach-Tarski paradox in similar terms from our discussion of Hilbert's paradox. Perhaps the most important distinction here is while our dicussion of Hilbert's hotel relied on only cardinality, the core of the Banach-Tarski paradox is *measure*, which we will discuss later.

Consider the set of all natural numbers:

$$\mathbb{N} = \{1, 2, 3, \cdots\}$$

We can divide this set up into two equally-sized sets (both countably infinite), by separating even and odd numbers. We already saw that both these sets are equivalent through the method of one-to-one correspondence we looked at in the earlier part. Since the sphere we are dealing with is not an actual sphere of atoms but a mathematical one defined as an infinite collections of points, we have a similar analogy. Voila! But, there's a much richer paradox behind this. Let's go into it in more depth.

12.2 A LITTLE BACKGROUND

12.2.1 *History*

The Banach-Tarski paradox (BTP) was introduced by Stefan Banach and Alfred Tarski, building off previous work from the likes of Giuseppe Vitali and Felix Hausdorff, both of whom were prominent analysts[1].

---

1 Mathematical analysis is a subfield of study which is concerned with the objects that shape calculus, including limits, derivatives, integrals, series, etc. This field grew from early calculus and built a more solid, rigorous structure for calculus to be studied under.

Figure 43: How did the missing square get filled in?

There have been many "paradoxes," much closer to visual illusions in their substance, that attempt to produce some sort of effect similar to BTP. Some of them were even viral internet sensations! These examples are simply geometric sleight of hand. Take for example the missing square illusion:

This is much more compelling, and surprising, when done in video-form, but the result is mind-blowing! It is indeed geometric sleight of hand here — the key to the puzzle is the fact that neither of the "triangles" above is truly a triangle because the hypotenuse is actually bent. The "hypotenuse" does not maintain a consistent slope, even though it definitely appears to do so. Another example was the viral internet sensation of "unlimited chocolate," in which chocolate bars were cut and reassembled into seemingly equal pieces but with each repetition of the process a new chocolate piece was removed.

The area is preserved in both of the above cases. As we have seen with the triangle, this is simply geometric sleight of hand! In the case of the chocolate bar, the pieces that were cut through actually get shorter, but the animation masks that in such a way to be almost impossible to realize. At the end of the day, we do have conservation laws in physics.

But we don't in math! So maybe the abstract triangle example is more than an illusion? Unfortunately not. Once again, we see that some visual illusion, namely the fact that our "triangles" are not truly triangles, is the driver behind these "paradoxes," and therefore it wouldn't be right to call them paradoxes at all.

On the other hand, BTP doesn't have any illusions. If anything, it's extremely hard to visualize! Mathematically, however, we can study it. And we are much more confident in our mathematical tools than our unassisted eye[2].

12.2.2 *How is This Theorem Unique?*

When I first heard of this theorem, I was already very familiar of the basic notions of set theory, transfinite numbers, and the like. I immediately thought of something similar to the analogy in the first section of this chapter and was unimpressed.

Even though that analogy provided a lens which I understood the paradox through, it masked the paradox's intricacies and obscured its real beauty and uniqueness. After further inspection, the paradox differed from those that I already wrapped my head around such as Hilbert's hotel. The result given by Banach and Tarski was something *much more* profound.

At first, I realized that the paradox's basis is rooted in measure, not the cardinality I was accustomed to seeing. While cardinality concerns discrete counting of elements, measure is concerned with the continuous.

Maybe, this is unique I thought to myself. But then again, I found some seemingly similar paradoxes regarding measure. Take for example a simple line.

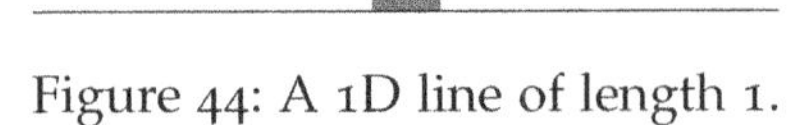

Figure 44: A 1D line of length 1.

2 Chapter 1 with all its visual illusions should serve as a reminder!

Now, abstractly speaking, a line is an uncountably infinite set of points. Let's see if we can manipulate these points to make two copies of our line.

It turns out, we can "project" each point from our line using a mapping like the one below. It is similar to how a beam of light spreads the further it propagates from its source.

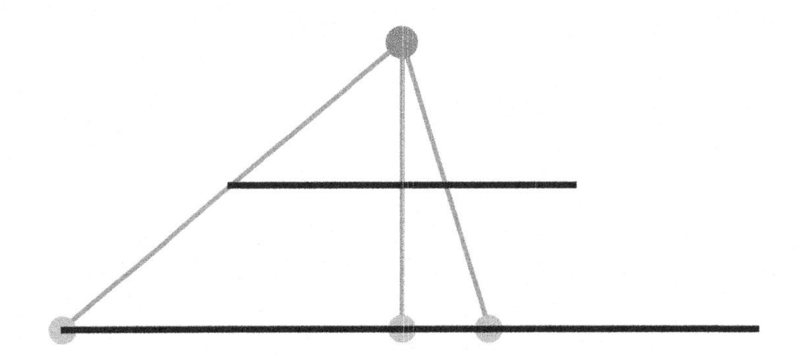

Figure 45: Points from line 1 being projected into line 2.

Simply mapping the points can double our line's length! As a matter of fact, we can map our line of length 1 to any larger length we desire using some basic knowledge of geometry and similar triangles[3]. And it gets even weirder than this.

In the previous part, we discussed how the cardinality of the set of all real numbers is an uncountable infinity. And if we take any interval $[a, b]$ such that $b > a$ on the real line, we end up with the same uncountable infinity! We can see this concept in direct application here, as the number of points making up line 1 and 2 in 45 is equivalent.

But what about a line and a solid square? Do they contain the same number of points? As it turns out, the number of

3 Notice that if you draw lines from our top point to the ends of the bottom line, you can see two similar triangles. Applying the rules of similarity can let us construct a point such that we map our line of length 1 to a line of arbitrary length greater than one.

points making up both objects is also the same! $\aleph_1$ to be exact. Consider the plot of the square below.

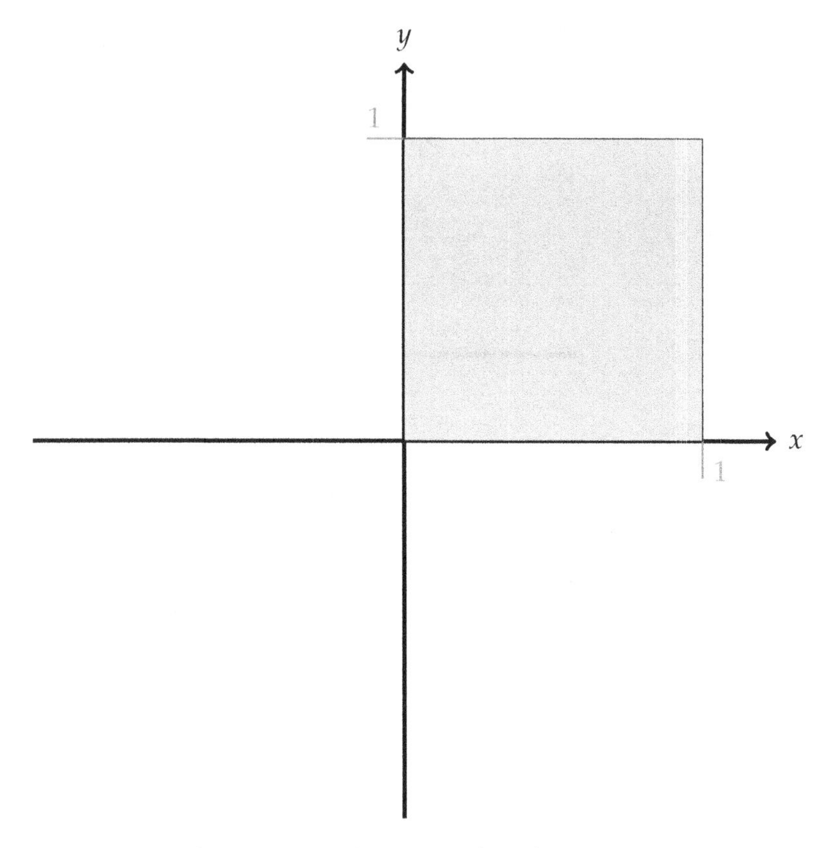

Figure 46: A Cartesian plot of a square.

Notice that the square above can be defined as the collection of all points (x, y) such that x and y are both in the interval $[0, 1]$. A hacky way to map the points on our line into those on the square is by thinking of "weaving" the coordinates of each point (x, y). For example, the point (.255, .456) will be mapped into a position of .245556 on our line. By doing this on all points that make up the square, we end up with a line that is exactly 1 unit in length! Subtracting no points, we mapped a square into a line. That is one weird result.

After the mathematician Dedekind introduced this observation to Cantor, Cantor, arguably the father of all of these paradoxes on infinity, replied

> *I see it, but I don't believe it.*

So don't worry if you find it hard to believe initially. The logic extends beyond a square to any 2D shape, and while we're on it, any shape in an arbitrary number of dimensions. It turns out the number of points in any $n-$dimensional shape (where $n \geq 2$) of any size is the same: $\aleph_1$.

But, the BTP is a bit different. It is not simply making use of the properties of the uncountably infinite and mapping the points. There's a bit of "rotation" involved, and to get at the heart of the rotation, we must understand the *hyper-webster*.

12.2.3 *Hyper-webster*

Consider a *hyper-dictionary* which contains every possible word the English language can assemble. This dictionary would contain every single possible sequence of letters, regardless of whether the word formed has any established meaning. The question is: what would be the best way to arrange our infinite dictionary?

The chief editor of the dictionary suggests 26 volumes, one for each starting letter. But, one of the assistant editors was quick to point out another way, saying "No! We can fit all words into one single volume." She suggests that they can fit all of the volumes' worth of words into one volume — volume A. Her reasoning is that since it is volume A, the reader would know to just strike the "A," the first letter of every word. By doing so, they have access to all the words in the hyper-webster! The reader has mathematically hacked his way into saving 25 volumes' worth of cash. We will see how this subtle character play helps us later in understanding this paradox. Enough with this discrete stuff, though, let's dive into more continuous things.

12.2.4 *Measure*

Consider a circle of radius 1. Now, remove one point from the circumference of that circle.

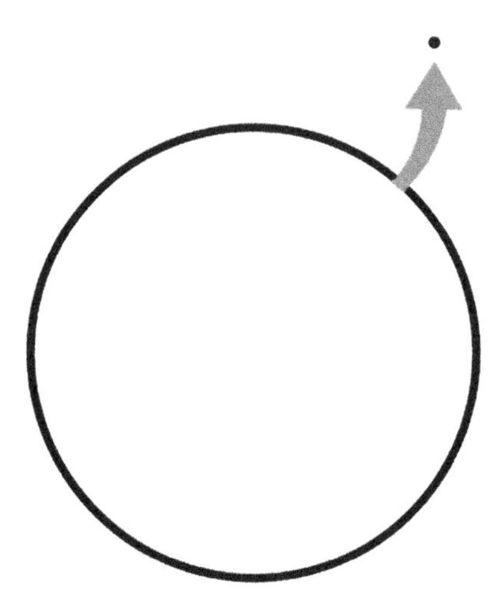

Figure 47: A circle with one point removed from its circumference.

Can you move the points in the circle in such a way as to fill that point? It turns out the answer is yes! Consider labelling the points on the circumference, starting from a random "starting point" and then labelling a new point everytime you cross a distance of 1 around the circle. Similar to the case of Hilbert's hotel, we can move point 1 to the empty point, point 2 to point 1, point 3 to point 2, and so on. After doing this, no empty points are left because they are never repeated! Weird.

The concept of measure is a necessity in any mathematical system. After all, mathematics in the form we know it today, which is based on proofs and deductive reasoning, was originally the study of geometric shapes through the likes of Pythagoras, Euclid, and others. But, once calculus was invented, our "obvious" concept became much less obvious. We needed a stronger and more rigorous concept to base calculus on. And mathematicians struggled for centuries with this.

That is, until Henri Lebesgue and others revolutionized the concept of measure.

In my previous book, *Advanced Calculus Explored*, I discuss how the notion of measure can be useful in the case of integrals that diverge at end points. The rigorous theorem behind this is named the *Lebesgue-dominated convergence theorem*: a very powerful result in higher mathematics which is useful to many scientific disciplines. One important phrase comes out of the theorem's statement: *almost everywhere.*

There is no tricky terminology here: "almost everywhere" is just what it means. More rigorously, a property holds *almost everywhere* on an interval if the set of all instances which the property does not hold has measure zero. Now, measure can be defined in many ways, but for the purposes of my last textbook, it was the Lebesgue measure. The Lebesgue measure very much resembles our common notions of measurement: the measure of a 1D line is its length, a 2D plane its area, a 3D blob its volume, and so on. By removing that single point from our continuous circle, the measure of its circumference is still the same.

12.3 A SIMPLIFIED EXPLANATION

Back to the BTP. After exploring some of the background concepts, it's now time to tackle it directly.

Let's start by considering a sphere, simply an *uncountable* collection of points ordered in such a way to produce a ball-shape. Now, we will try to name each point on the surface of this sphere by assigning each point to a sequence of moves to get there from a starting point, each of length $\arccos\left(\frac{1}{3}\right)$ to ensure that no points get counted more than once. For example, we can call point A in figure 48 "UR," after the sequence of events "Right, Up." Notice that we apply moves from *right to left*, which we will see the importance of later on.

We have four total moves we can do: Left, Right, Up, and Down (L, R, U, D, respectively). Of course, instances of LR, RL, UD, or DU are not permissible as they will cancel out. We can then imagine listing each possible sequence of moves and coloring points based on the last move we make. Similar to a hyper-webster, but with some restrictions.

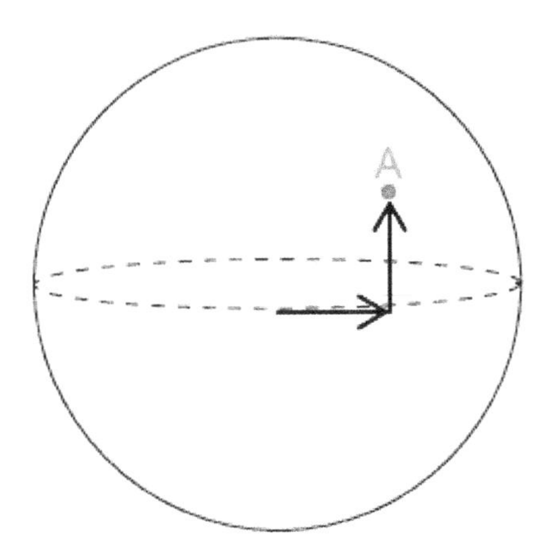

Figure 48: Point A on our sphere, which can be accessed by the sequence UR.

However, starting with only one point is inevitably going to make us miss some points. To avoid this, we will select all points that we miss to be starting points too (which we will color distinctly). Alongside our hyper-webster of moves, this would name every point. Except for *poles*.

The problem with these poles is that they can be counted and named more than once. Adding any subsequent moves can preserve the point's location. We know that every sequence has two, so we will just color them another color. We now have *six* total decompositions of the sphere's surface: starting points, poles, and a decomposition for each move (L, R, U, D). Since we are dealing with a solid sphere, we can extend each point into a line ending at the center of the sphere, making seven total pieces with the center (it is not included in any of the other pieces). Now comes the *really* neat part.

Take the "left" piece. Based on our definition, it is simply the set of all points on the surface that had a movement sequence ending with a "left," i.e.

$$L, LL, \cdots, LU, LUU, \cdots, LD, LDD, \cdots, LUL, LULL, \cdots$$

Let's rotate this piece to the right. This is just an additional R at the end of the sequence, which we write as the first letter (remember, we perform sequences from right to left). But if we add the R to our sequences, we can cancel the first L to get:

$$, L, \cdots, U, UU, \cdots, D, DD, \cdots UL, ULL, \cdots$$

We now have every move that ends with L, U, D, and a blank move in the beginning, which is equivalent to all starting points! We now have turned our left piece into four pieces with simple character play. Just like the hyper-webster!

Adding the right piece, poles, and center to our rotated left piece, we have ourselves a whole new sphere! But, there are still three pieces left over: the up, down, and starting points pieces. Now, we will try to rotate these pieces in such a way to make a new sphere. Let's take a look at the "Up" piece. For this piece, however, instead of rotating all points, we will rotate all points that aren't a pure string of U's.

$$U, UU, UUU, \cdots, UL, ULL, \cdots, UR, URR, \cdots, ULU, ULUU, \cdots$$

Turns into

$$U, UU, UUU, \cdots, L, LL, \cdots, R, RR, \cdots, LU, LUU, \cdots$$

Which gives us the Up, Left, and Right pieces! We will now combine this rotated Up piece with the down pieces and starting points to get a whole new sphere. That is, except the poles and the center point. To fix this, recall the circle with the missing point. The number of holes in the sphere is countably infinite, being made up of just the poles and the center point. We can fill them all just like we did in the example of the circle.

Wow, that was a marathon! We just took one ball and somehow transformed it into two equivalently-sized balls. Almost seems like magic! The implications of this for both abstract mathematics and science can be huge. The paradox relies critically on the *axiom of choice*, an initially controversial axiom of set theory, and provokes thought into the nature of mathematics. The axiom of choice is an axiom of set theory equivalent to the statement that the *Cartesian product of a collection of non-empty sets is non-empty*. Informally, it states that given any collection of bins, each containing at least one object, it is possible to make a selection of exactly one object from each bin, even if the collection is infinite. Initially, this axiom was highly controversial because some of the consequences of the axiom, like the BTP and others, were highly paradoxical.

Okay, it might be important mathematically, but could this paradox have any physical meaning?

12.4 REAL WORLD EXAMPLES?

If matter were infinitely divisible, which it is not according to our most cutting-edge theories, then it might be possible. Even then, the pieces we defined are so exotic that they don't have a well-defined notion of volume (or more generally, measure) associated with them. As much as the Banach-Tarski paradox is about weird set-theoretic geometry, it is about fundamental notions we have about mathematics and breaking them down. Paradoxes like this one perhaps best embody mathematical exploration because they invite us to lands that challenge our current knowledge to the highest degree.

There has been increasing interest in the application of this paradox to particle physics, particularly the manner in which particles collide. Hadronic physics[4], specifically, shows this

4 Hadrons are simply subatomic composite particles made up of two or more quarks. Most of the mass in ordinary matter comes from two hadrons: the proton and the neutron.

resemblance. The details are a bit advanced, and can be found in Augenstein's paper "Hadron physics and transfinite set theory."

Throughout history, there have been countless examples of things that "didn't make sense" that eventually turned into the most useful science we know. From negative numbers to imaginary numbers, the abstraction of many mathematical concepts has always caused controversy. Nonetheless, these numbers find application in almost every scientific field now. Beautifully, mathematics often finds its way into theories of the physical world within a few centuries. Could the Banach-Tarski paradox be just another example of that?

Part V

MIND-BENDING STATISTICS

All models are wrong, but some are useful.
— George E. P. Box

Statistics shapes our lives. From analyzing census data to determine where government funding should be directed to making scientific inferences, the study of *data* has become of ever-growing importance. Especially in the age of AI and machine learning, which is almost entirely based on statistical theory, studying statistical paradoxes is crucial.

These paradoxes are perhaps the least fundamental, but also arguably the most frequently encountered. They appear in medical and social studies all the time, and it is very important to be aware of them to make accurate conclusions.

13

BERKSON'S PARADOX

We will start this chapter off in a slightly different way. Why do attractive people seem meaner than average? Yes, seriously. Is that sentiment true? Or is there something unseen behind that intuition?

Our experiences and memories determine our judgements of people, events, ideas, and everything else we may have an opinion on. At root, we are subconsciously data-driven creatures. That is why we are so good at learning.

It's also why machine learning is powerful: data. Neural networks are especially powerful — the structure of our brain makes it excellent at digesting data, so neural network machine learning algorithms are modeled after it. In the end, it is clear that data is power! So, back to the matter at hand, are attractive people meaner than average?

A popular Slate article, "Why Are Handsome Men Such Jerks?", was written on this subject by mathematician Jordan Ellenberg in 2014. We will discover that the key lies in the subtle erroneous sample selection we use as humans in our personal relationships, which eventually leads to Berkson's paradox[1]. However, this phenomenon extends far beyond attractive jerks — it actually started out with something that was much more dangerous to form incorrect conclusions about: medicine.

1 More on this in a couple of sections!

13.1 MEDICAL DATA

Berkson's paradox is named after American medical statistician Joseph Berkson, who first encountered and explained it through his medical studies.

Joseph Berkson was trained as a physicist, doctor, and statistician. He was head of the Division of Biometry and Medical Statistics at the Mayo Clinic, a leading medical research institution, for 30 years. In addition to his paradox, he is known for being a leading biostastician, publishing over a hundred scientific papers.

Figure 49: Joseph Berkson (1899 – 1982)

Berkson explained the effect using a past study on a risk factor for a disease. The statistical sample in the study was from a hospital in-patient population.

Because the statistical sample included only the hospital in-patient population, rather than a random sample of the general public, the results of the study can indicate a fictitious negative association between the disease and the risk factor.

In Berkson's original paper, "Limitations of the Application of Fourfold Table Analysis to Hospital Data," he discusses two totally unrelated diseases that hospital data implied to be connected somehow. Berkson used the example of cholecystitis, or the inflammation of the gallbladder, and diabetes. At the time, there was a widespread belief in the medical community that cholecystitis is a risk factor for diabetes. In some places, the gallbladder was even being removed in an attempt to treat diabetes! As we can see, statistics can *literally* be a life-changer, but we have to use it carefully to ensure that life change is for the better.

A hospital patient without diabetes is more likely to have cholecystitis than an individual from the general population, as the patient must have had some non-diabetes — possibly cholecystitis causing — reason to enter the hospital. Then, we see that the connection between diabetes and cholecystitis will be obtained regardless of any association between diabetes and cholecystitis in the general population. Subtle sample selection mistakes can be dangerous!

The end result is a classic tale of "correlation does not imply causation." In this case, it was a bad case of ascertainment, or sampling, bias, in which the statistical sample is not appropriate for the study.

13.2 ARE ATTRACTIVE PEOPLE JERKS?

Let's go back to our initial question:

Are attractive people jerks?

But now, we are armed with a fuller understanding of statistics and data — so we won't make the same mistakes. With Berkson's paradox in mind, let's get crunching!

Let's start with a reasonable assumption: attractiveness and (the lack of) kindness are virtually independent traits. So, we suspect that if we sample a random population of 5000 people, we would have a distribution that looks something like that in figure 50 (shown in two pages).

But, let's be honest, your personal sample size is probably restrained. If we divide this data up, we will have four types of people: unkind and unattractive, kind but unattractive, unkind but attractive, and kind and attractive. Most people would never even consider those who are both unkind and unattractive! We are now starting to see where sample size might be restricted.

Let's suppose that there is a minimum threshold for Kindness + Attractiveness for a person to consider another person romantically. Each variable is on a scale from 0-10, with 10 being the highest possible rating. Let's be mildly picky and set that threshold to 15, i.e. for you to consider a person romantically,

$$\text{Kindness} + \text{Attractiveness} > 15$$

We can graph this threshold on figure 50 to see how it impacts our data. We will use Mathematica, as we previously did with our other plots, for this.

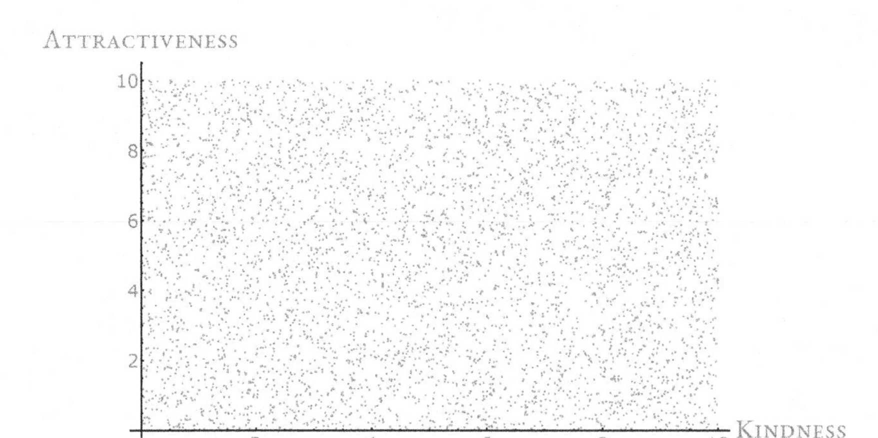

Figure 50: Each point indicates a person's rating. As we can see, there is no correlation between kindness and attractiveness here.

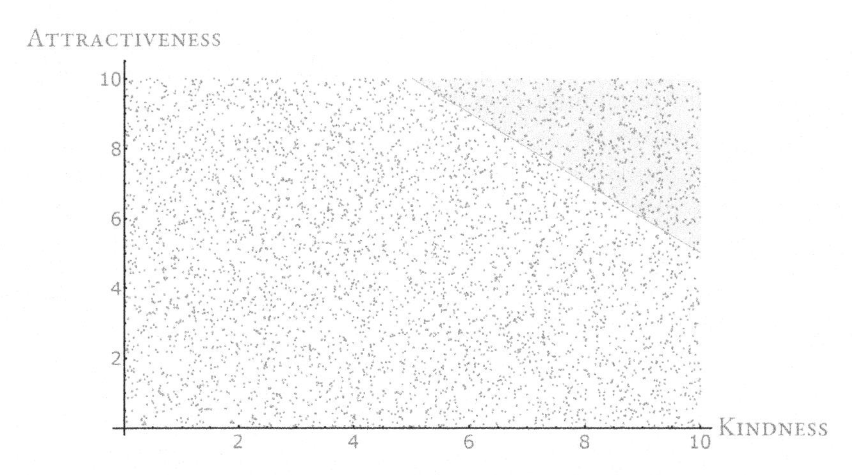

Figure 51: The highlighted area shows the region one considers romantically.

Wow, maybe that isn't mildly picky after all! If we calculate the area of that triangle up-top, we get 12.5 out of a total area of 100. Assuming equal distribution, that's the top 12.5%!

Anyways, we can detect a correlation here. If we try to find a linear correlation, we get the equation:

$$\text{Attractiveness} = 12.69 - 0.52 \cdot \text{Kindness}$$

Giving a negative correlation between kindness and attractiveness!

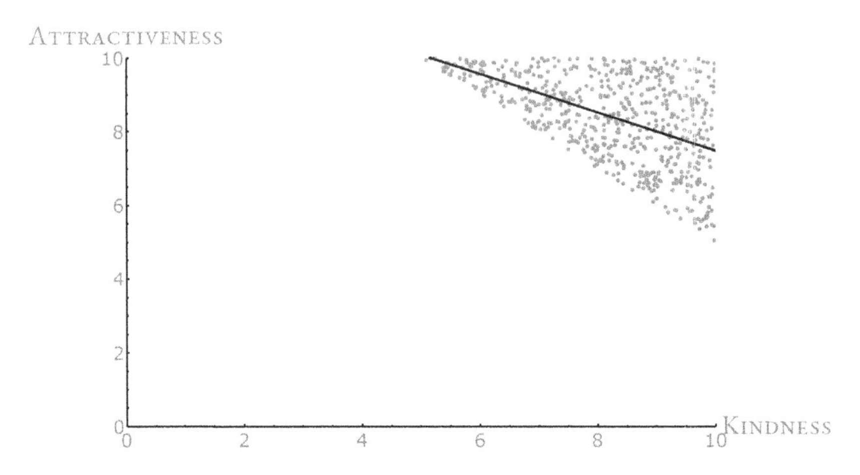

Figure 52: The fitted linear equation (13.2) illustrated in black versus the randomly generated sample data in blue

The higher threshold you set, the more the trend seems to hold up! Let's have some fun and try a threshold of say, 18. This is only looking at 2% of the total population.

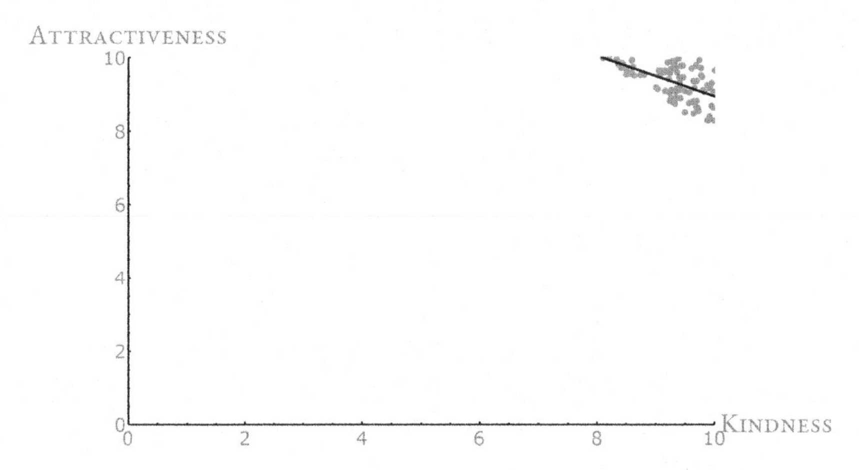

Figure 53: We have a slightly stronger negative correlation here, with the slope being less than $-.54$.

Compare this with a non-picky person, with a threshold of 10. This is considering 50% of the population.

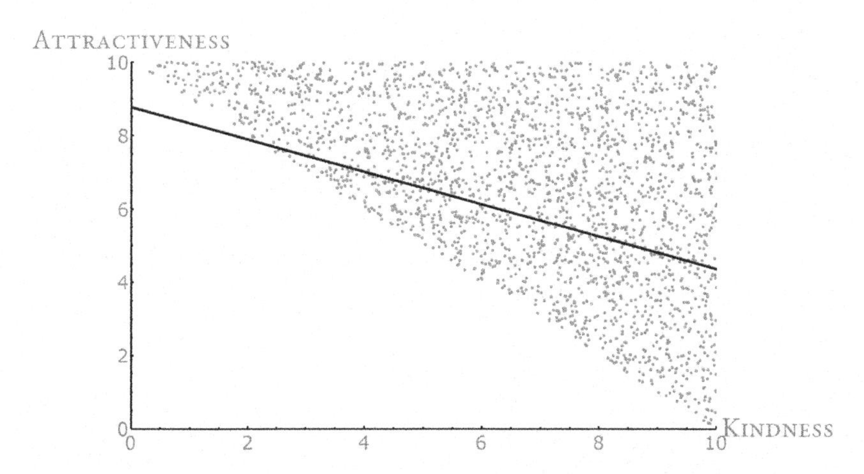

Figure 54: We have a slightly weaker negative correlation here, with the slope being greater than $-.45$. Also, the correlation coefficient is decreasing! It turns out the less biased we are in our sample, the less potent our fictitious conclusions are.

Of course, our brain is not crunching up numbers and running them through a linear regression algorithm. Nonetheless, our brain is constantly trying to make conclusions on the data that we have. And here we see the mishaps of statistics and our data-driven minds!

13.3 MORE EXAMPLES

We could delve into even more examples of this phenomenon, as it is more common than it seems. It can explain away many false conclusions we draw from everyday experiences, and we'll cover just one here.

13.3.1 *Does Hollywood Ruin Film Adaptations?*

Does it? It seems that every novel Hollywood touches, it makes a poorly adapted film out of it! But is that really true, or is Berkson's paradox at play here?

As it turns out, it is indeed Berkson's paradox in action. Thinking alongside the lines of our previous question about attractive people and their kindness, or lack thereof, we can again try to assume that there is no correlation between the quality of a movie and the quality of a book. It is presumably a bit harder to make a good movie out of a bad book, but do not worry, this will not impact our final result. The reason is our limited sample, which will force us to overwhelmingly consume movies in the top right corner.

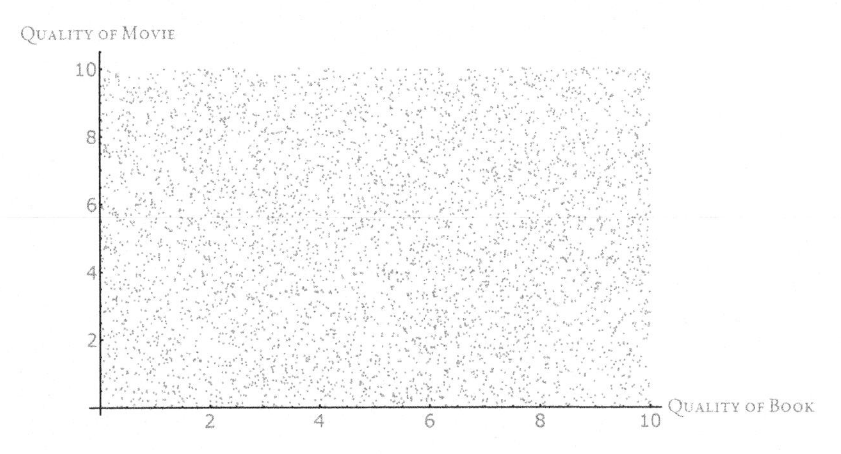

Figure 55: This figure was generated by plotting 5000 points randomly.

Really, you are only looking at a small subset of the points here. Either a movie was so bad you did not even hear about it, or a book was so bad you did not think to look up its movie adaptation. The result is a sample very much similar to that we looked at previously where you are only really looking at the top right corner of the data. And in here we can see why that apparent negative correlation between quality of book and quality of movie exists.

14

SIMPSON'S PARADOX

Simpson's paradox is another example of how paradoxes have helped shape the field of statistics. If statistics was so intuitive, its study would be much less valuable! Simpson's paradox closely mirrors Berkson's paradox in that it also is a consequence of a badly chosen sample. However, Simpson's paradox differs in that when an actual correlation is present, a badly chosen sample can lead to the opposite correlation. The effect was first described by Edward H. Simpson in 1951. However, similar effects were mentioned earlier in the work of statisticians Udny Yule and Karl Pearson.

Edward Hugh Simpson was a British mathematician and code-breaker. His paradox was featured in various shows, including the popular American sitcom "The Simpsons"![a]

[a] That show never disappoints with their creativity.

Figure 56: Edward Simpson (1922 - 2019)

In essence, the paradox is about the simple arithmetic fact that

$$\frac{a}{b} > \frac{c}{d} \text{ and } \frac{A}{B} > \frac{C}{D}$$

Does not imply

$$\frac{a+A}{b+B} > \frac{c+C}{d+D}$$

14.1 COLLEGE ADMISSION SCANDAL?

An interesting case of this paradox in action comes from a lawsuit against the University of California at Berkeley (UC Berkeley). Due to 1973 admission data, UC Berkeley was sued for discriminating against women who were applying to their graduate programs!

The data indicated that, overall, men were accepted at a higher rate than women at a statistically significant rate — a red flag for possible discrimination. However, with more careful statistical study — and the knowledge of Simpson's paradox — it was revealed that there was no bias against women. If anything, a slight bias against men was found!

The rough explanation for this is women tended to apply for more selective graduate programs than their male counterparts. To get more of an understanding behind this, let's bring out some data in table 3. This data is from the six largest departments.

If one considers each department separately, women are actually admitted at higher rates in the majority of departments (4/6). This is indicated through the emboldened numbers. The departments with the highest number of male applicants, 1 and 2, are the least selective, while those with the most female applicants, 3 and 5, have rather low admission rates.

Nonetheless, when one first considers the overall data, it appears to be very apparent there is discrimination! Out of 8442 male applicants, 44% are accepted, while out of 4321 female applicants, 35% are accepted.

When first looking at this, it seemed obvious that this disparity must have been due to discrimination. But, when looking at it through a more careful lens, this conclusion proves out to be false. For more, you can read into the original research paper by Bickel et Al. — "Sex Bias in Graduate Admissions: Data From Berkeley."

Department	Men		Women	
	Applicants	Admitted	Applicants	Admitted
1	825	62%	108	**82%**
2	560	63%	25	**68%**
3	325	**37%**	593	34%
4	417	33%	375	**35%**
5	191	**28%**	393	24%
6	373	6%	341	**7%**

Table 3: Admission data from 1973 at UC Berkeley.

14.2 VISUALIZATION IS KEY

Essentially, the problem here is one of drawing inferences from a limited sample size. The challenge is: how can we make decisions based on data that gives out different conclusions if partitioned? Kievit et Al. (2013) argue that Simpson's paradox may occur in a wide variety of statistics, particularly within the social and medical sciences. The researchers stress the importance of data visualization in helping recognize patterns that might not be obvious through looking at mere numbers. This may be easy if your data looks like this:

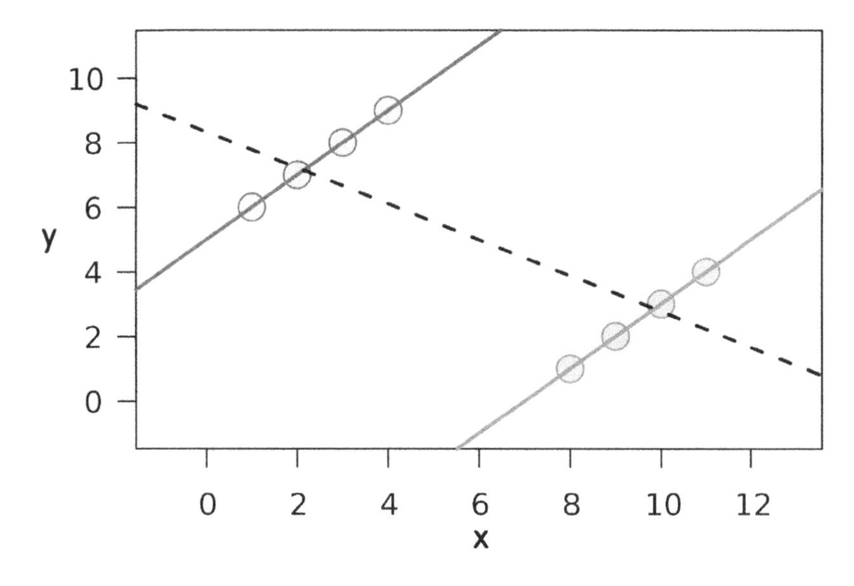

Figure 57: Visual representation of Simpson's paradox

In figure 57, one can easily see a positive trend emerging in both the pink and blue points, whereas when one combines the two groups the opposite trend is seen. This is essentially what is at play here — merely an issue of sample selection. To treat statistics as a game of only numbers would be naive. A nice example of the downfalls of this approach is given by Justin Matejka and George Fitzmaurice in their 2017 research paper: "Same Stats, Different Graphs." In it, they expand on some work that is almost 50 years old: Anscombe's Quartet. Anscombe's Quartet is a classical idea in statistics, and is a group of four datasets created by the English statistician Frank Anscombe in 1973 that have the same "summary statistics," i.e. mean, standard deviation, and Pearson's correlation. Yet, they all have wildly different graphs!

Matejka and Fitzmaurice took this idea and ran with it. They generated a dozen wildly different graphs that somehow have the same summary statistics to two decimal places.

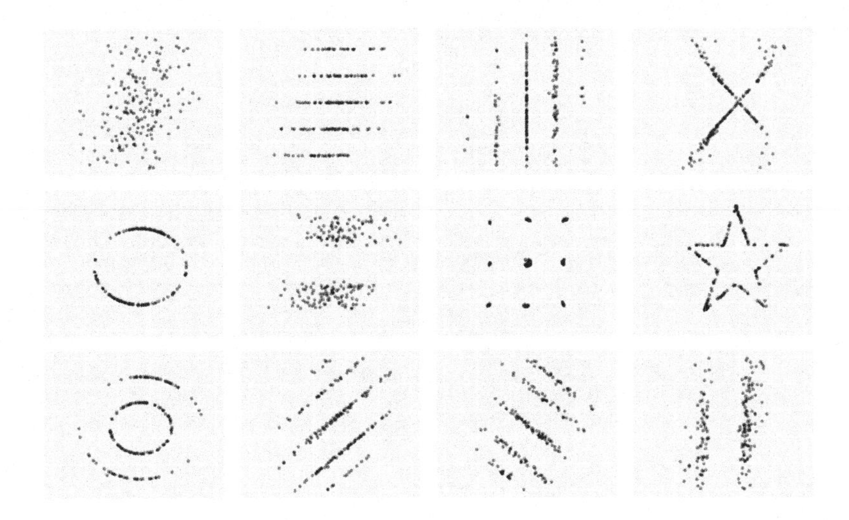

Figure 58: The dozen graphs Matejka and Fitzmaurice generated.

In their paper, they also mention Simpson's paradox! The point here is, data visualization is more than just an aesthetic addition. It is a crucial part of understanding and doing statistics.

Here, we must be reminded that statistics is both an art and a science. We have our statistical tools — regression models, algorithms, etc — akin to an artist's toolkit made up of brushes and paints. That is our science. But the art of statistics is applying these tools to construct useful models we can draw correct conclusions from. After all, what is an artist if they do not make art with their tools! Both of the paradoxes we discussed in this part are falsidical paradoxes, which then again are excellent educational tools. Through these two paradoxes, one can learn many lessons about statistics!

15

A SEND-OFF

Wow, what a ride! From ancient examples like Zeno's paradoxes to modern, advanced paradoxes such as the Banach-Tarski paradox, we explored some of the most significant paradoxes in the history of mathematics.

We have seen how paradoxes have guided mathematical progress for millennia. They showed us what truth to look for through fundamental paradoxes like Russell's paradox and how to look for truth through confusing objects like Grandi's series. Ultimately, though, this book's purpose extends far beyond its mathematical content. There is a reason philosophy and history were omnipresent here — because as much as paradoxes teach us about mathematics, they teach us much about the history of philosophy and the development of rational thought as we know it today.

The application of the paradoxes discussed in this book knows no bound. Paradoxes like Zeno's paradox helped us understand and better formulate the ideas of calculus, which drives virtually all of science. Moreover, statistical paradoxes like Berkson's and Simpson's paradox have resulted in more refined and statistically-valid medical and social studies, thereby guiding decision-making to benefit society immensely. It would be difficult to find a part of life that these paradoxes have not touched!

It's worth remembering the reason why these paradoxes are important. They're fun, yes! But they are much more than brain-teasers; they are driving forces behind mathematical progress and invaluable tools to advance mathematics education. It's time to realize the importance of these paradoxes, and incorporate them into our school curriculums more so than ever.

The world has enough calculators — we have billions of computers that can out-calculate any human. But, we will never be in a surplus of problem solvers. Paradoxes teach problem-solving in a beautiful and insightful way. They teach us to escape our intellectual comfort zone in pursuit of truth. The earlier we realize this, the better armed the next generation of mathematicians, scientists, engineers, and many others will be. Let's give the next generation of thinkers the tools they need to add value in a world full of calculators. See you in the next book of the *Paradoxes* series!

Part VI

APPENDIX

A BRIEF INTRODUCTION TO CALCULUS

We talked a bit about calculus in this book, so let's discuss some of its basics. Calculus is simply the study of change — as well as a lot of other related concepts. There are two main branches of calculus — differential and integral. These concern derivatives and integrals, respectively, which are essentially opposite operations.

Derivatives give you an idea of the "slope," or the rate of change, of a certain function. They let us plug in a certain value and get a value for the instantaneous rate of change in the original function back. They are usually indicated by the operator $\frac{\mathrm{d}}{\mathrm{d}x}$, where the notation $\mathrm{d}x$ indicates the variable we're interested in measuring the rate of change against is x. If we have a function, $f(x)$, for example, we can also denote its derivative by adding a 'prime' as a superscript, i.e.:

$$\frac{\mathrm{d}}{\mathrm{d}x}f(x) = f'(x)$$

On the other hand, there are two types of integrals — definite and indefinite. An indefinite integral is simply defined as the "anti-derivative" of a certain function, which is basically asking the question "What function, when we take its derivative, gives us our current function?" A definite integral has bounds, however, and basically gives us the "area under the curve" between these bounds.

All of these concepts are related by the *fundamental theorem of calculus,* which has two parts. The first part essentially shows that differentiation and integration are opposite operations. The second part shows that the *definite integral* is simply the result of subtracting the values of the indefinite integral at the bounds. For example, say we have a function $f(x)$ whose antiderivative is $F(x)$. We are trying to find the area under $f(x)$ from $x = a$ to $x = b$, then:

$$\int_a^b f(x)\,\mathrm{d}x = F(b) - F(a).$$

The $\mathrm{d}x$ indicates that we are integrating *with respect to* x, i.e. it is our variable of our interest. Of course, the underlying concept behind all of these objects is the *limit,* which can be essentially explained as the question "What value does my function approach as I am approaching a certain x−value?" Although it might seem like a useless concept, as we saw when we looked into the definition of the derivative in chapter three, it is a very useful concept.

B

A BRIEF INTRODUCTION TO SET THEORY

A set is simply a collection of items. Any type of items. Usually, these items are numbers. You can write a set by listing these items (called the set's 'elements') inside curly braces. For example, say the set A has the elements 1, 2, and 3. This can be expressed as:

$$A = \{1, 2, 3\}.$$

The order of the elements does not matter in a set, but it is conventional to write the elements of a set in ascending order (from smallest to largest). Also, it does not matter if the set lists the same element multiple times — the result is identical to a list where that element appears only once.

C

NOTATION GUIDE

Now, onto some notation.

Symbol	Meaning
$\frac{\mathrm{d}}{\mathrm{d}x}$	The derivative of a function with respect to x
$f'(x)$	The derivative of $f(x)$
$\int f(x)\,\mathrm{d}x$	The integral, or anti-derivative, of $f(x)$ with respect to x
$\int_a^b f(x)\,\mathrm{d}x$	The area under $f(x)$ between a and b
$\lim_{x\to a} f(x)$	The limit of $f(x)$ as x approaches a

Table 4: Basic calculus notation

Symbol	Meaning
$\cup$	The union of two sets
$\cap$	The intersection of two sets
P(A)	The power set, or the set of all subsets, of A
$a \in A$	a is an element of A
$a \notin A$	a is not an element of A
$\aleph_0$	The smallest cardinal number, aleph zero
$\aleph_1$	The second smallest cardinal number, aleph one
$\mathbb{N}$	The set of all natural numbers
$\mathbb{Q}$	The set of all rational numbers
$\mathbb{R}$	The set of all real numbers

Table 5: Basic set theory notation

There are, of course, countless other symbols in these fields of study, but these shall do for this book.

D

ACKNOWLEDGEMENTS

I would like to sincerely thank my friends and family, who have made the long journey of writing this book enjoyable. With their company, I fought against many hurdles.

I am also grateful to those who have helped polish this book into what it is now. I am especially indebted to three members of the DailyMath community:

- Duc Van Khanh Tran
- Kevin Tong
- Keyvon Rashidi

for their feedback on the first draft of the book. I am also thankful for my editor, Rick Keys from BookOneDone, who was instrumental in the post-writing process.

Oh, I almost forgot, a special thank you goes out to the DailyMath community at large!

The references on the next page are arranged in alphabetical order based on the author's last name.

REFERENCES AND FURTHER READING

Wazir H. Abdi. *Toils and Triumphs of Srinivasa Ramanujan: The Man and the Mathematician*. National Publishing House, 1996.

Hamza E. Alsamraee. *Advanced Calculus Explored: With Applications in Physics, Chemistry, and Beyond*. Curious Math Publications, 2019.

B. W. Augenstein. Hadron physics and transfinite set theory. *International Journal of Theoretical Physics*, 23(12):1197–1205, 1984.

Giorgio T. Bagni. Infinite series from history to mathematics education. *International Journal for Mathematics Teaching and Learning*, Jun 2006.

E.J. Barbeau and P.J. Leah. Euler's 1760 paper on divergent series. *Historia Mathematica*, 3(2):141–160, 1976.

Joseph Berkson. Limitations of the application of fourfold table analysis to hospital data. *Biometrics Bulletin*, page 47–53, Jun 1946.

Bruce C. Berndt. *Ramanujan's Notebooks*. Springer, 1989.

P. J. Bickel, E. A. Hammel, and J. W. Oconnell. Sex bias in graduate admissions: Data from berkeley. *Science*, 187(4175):398–404, 1975.

Carl B. Boyer. *The History of the Calculus and its Conceptual Development*. Dover Publications, 2012.

Bryan Hammond Bunch. *Mathematical Fallacies and Paradoxes*. Dover Publications, 1997.

William Byers. *How Mathematicians Think: Using Ambiguity, Contradiction, and Paradox to Create Mathematics*. Princeton University Press, 2010.

Eugenia Cheng. *Beyond Infinity: An Expedition to the Outer Limits of Mathematics*. Basic Books, 2017.

Matt Cook. *Sleight of Mind: 75 Ingenious Paradoxes in Mathematics, Physics, and Philosophy*. MIT Press, 2020.

A. Einstein, B. Podolsky, and N. Rosen. Can quantum-mechanical description of physical reality be considered complete? *Physical Review*, 47(10):777–780, 1935.

Albert Einstein, Max Born, and Hedwig Born. *The Born-Einstein Letters: Correspondence between Albert Einstein and Max and Hedwig Born from 1916 to 1955*. Walker, 1971.

Jordan Ellenberg. Why are handsome men such jerks?, Jun 2014.

Euclid and Thomas Little Heath. *The Thirteen Books of Euclid's Elements*. Cambridge University Press, 2015.

Leonhard Euler. *Elements of Algebra*. Springer-Verlag, 1984.

Stanley J. Farlow. *Paradoxes in Mathematics*. Dover Publications, 2014.

Giovanni Ferraro and Marco Panza. Developing into series and returning from series: A note on the foundations of eighteenth-century analysis. *Historia Mathematica*, 30(1):17–46, 2003.

Richard P. Feynman, Robert B. Leighton, and Matthew L. Sands. *The Feynman Lectures on Physics*. Basic Books, 2011.

T. Fujii. The role of cognitive conflict in understanding mathematics. *Proceedings of the International Conference on the Psychology of Mathematics Education*, 1987.

George Gamow. *One, Two, Three ... Infinity: Facts and Speculations of Science*. Dover, 1988.

Ganesh Gopalakrishnan. *Computation Engineering: Applied Automata Theory and Logic*. Springer, 2011.

Judith V. Grabiner. Who gave you the epsilon? cauchy and the origins of rigorous calculus. *The American Mathematical Monthly*, page 185–194, Mar 1983.

John Green. *The Fault in Our Stars*. Dutton Books, 2018.

Brian R. Greene. *The Elegant Universe: Superstrings, Hidden Dimensions, and the Quest for the Ultimate Theory*. Jonathan Cape, 1999.

K. Irwin. What conflicts help students learn about decimals. *Proceedings of the Twentieth Annual Conference of the Mathematics Education Research Group of Australasia*, page 247–254, 1997.

Morris Kline. Euler and infinite series. *Mathematics Magazine*, 56(5):307, 1983.

Sergiy Klymchuk. *Counter-examples in Calculus*. Mathematical Association of America, 2010.

Sergiy Klymchuk. Using counter-examples in teaching and learning of calculus: Students' attitudes and performance. *Mathematics Teaching Research Journal Online*, 5(4), Jan 2012.

Justin Matejka and George Fitzmaurice. Same stats, different graphs. *Proceedings of the 2017 CHI Conference on Human Factors in Computing Systems*, 2017.

Adrian William Moore. *The Infinite*. Routledge, third edition, 2019.

Ernest Nagel, James R. Newman, and Douglas R. Hofstadter. *Godel's Proof*. New York University Press, 2008.

Eugene P. Northrop. *Riddles in Mathematics: A Book of Paradoxes.* Dover Publications, 2014.

Ben Orlin. *Math with Bad Drawings: Illuminating the Ideas That Shape Our Reality.* Black Dog & Leventhal Publishers, 2018.

Matt Parker. *Things to Make and Do in the Fourth Dimension: A Mathematician's Journey Through Narcissistic Numbers, Optimal Dating Algorithms, at Least Two Kinds of Infinity, and More.* Penguin Books, 2015.

Bertrand Russell. *Introduction to Mathematical Philosophy.* Allen and Unwin, 1919.

Richard M. Sainsbury. *Paradoxes.* Cambridge University Press, 2013.

Wilfried Sieg. *Hilbert's Programs and Beyond.* Oxford University Press, 2013.

Edward H. Simpson. The interpretation of interaction in contingency tables. *Journal of the Royal Statistical Society: Series B*, 13(2):238–241, 1951.

Simon Singh. *The Simpsons and Their Mathematical Secrets.* Bloomsbury Publishing Ltd., 2013.

Ian Stewart. *Professor Stewart's Hoard of Mathematical Treasures.* Basic Books, 2010.

Steven H. Strogatz. *Infinite Powers: How Calculus Reveals the Secrets of the Universe.* Houghton Mifflin Harcourt, 2019.

James F. Thomson. Tasks and super-tasks. *Analysis*, 15(1):1–13, Oct 1954.

Leonard M. Wapner. *The Pea and the Sun: A Mathematical Paradox.* CRC Press, 2005.

Aristotle. *Physics.* Oxford University Press, 1996.

Anne Watson and John Mason. *Mathematics as a Constructive Activity: Learners Generating Examples*. Routledge, 2008.

Joan Weiner. *Frege in Perspective*. Cornell University Press, 2009.

Alfred North Whitehead and Bertrand Russell. *Principia Mathematica*. University Press, 1968.

Conrad Wolfram. *The Math(s) Fix: An Education Blueprint for the AI Age*. Wolfram Media, Inc., 2020.

INDEX

Made in the USA
Coppell, TX
19 November 2020